Simon Henry Gage

The Microscope and Histology

Simon Henry Gage

The Microscope and Histology

ISBN/EAN: 9783337710361

Printed in Europe, USA, Canada, Australia, Japan

Cover: Foto ©berggeist007 / pixelio.de

More available books at **www.hansebooks.com**

THE

MICROSCOPE AND HISTOLOGY

BY

SIMON HENRY GAGE,

Associate Professor of Physiology and Lecturer on Microscopical
Technology in Cornell University, Ithaca, N. Y., U. S. A.

PART I.

THE MICROSCOPE

AND

MICROSCOPICAL METHODS.

Illustrated with six plates and eight figures in the text.

FOURTH EDITION, REVISED AND ENLARGED.

PUBLISHED BY JAMES W. QUEEN & CO.,
1010 CHESTNUT STREET,
PHILADELPHIA, PA.
1892.

Copyright, 1892.
BY SIMON HENRY GAGE,
All Rights Reserved.

Printed by
ANDRUS & CHURCH
Ithaca, N. Y.

PREFACE TO THE FOURTH EDITION.

In the use of the third edition of this work with large classes and advanced students, the absence of illustrations for the chapter on the micro-spectroscope and micro-polariscope was found to be a serious defect. To remedy the defect an additional plate has been added to this edition; and with the plate are given several spectra to elucidate the text on the one hand, and on the other to depict the appearances actually seen in working with the micro-spectroscope more accurately than in Plate V. (See Plate VI and explanation opposite p. 54.)

It was found also that the plan and directions for arranging serial sections (p. 78) were not satisfactory in embryological work, they have, therefore, been replaced by those which are believed to be more logical and convenient, and of more general application.

The author is greatly indebted to Professor Wilder, to Instructors Hopkins and Fish for suggestions in devising the plan now offered.

Besides the changes indicated above, a few minor ones were found necessary as indicated in the table of corrections and additions on the reverse of this sheet.

The author would feel grateful to any person who uses this book, if he would point out any errors of statement that may be discovered, and also suggest modifications which would tend to increase the intelligibility, especially to beginners.

Finally, as indicated on the title page, the work is now published by Messrs. James W. Queen & Co.

SIMON HENRY GAGE.

MAY 20, 1892.

ADDITIONS AND CORRECTIONS.

Corrections: Table, p. 6. In the last catalog of Leitz, the tube-length is given as 160 mm., and this length is said to be measured between the points a–d of the diagram (Fig. 8^b.)

P. 7, eighth line from the bottom: For N.A. = sin n or 1 × sin u, etc., read N.A. = n or 1 × sin u.

P. 14, last two lines: For "the eye-point is nearer the eye-lens in low than in high oculars," read, The eye-point is usually nearer the eye-lens in high than in low oculars.

P. 30, § 77, second paragraph: Only part of the field will be lighted if a low power is used and a very small diaphragm is close to the object; but if the small diaphragm is considerably below the object, the whole field will be lighted, although not very satisfactorily.

P. 38, paragraphs 1 and 3: For "Wallaston" read Wollaston.

P. 40, third paragraph: For "presbyopic" read hypermetropic.

P. 48, § 121, first paragraph, next the last line: For "party" read partly.

P. 54, § 128, first paragraph, second line from the end: For "emergent rays parallel" read emergent rays approximately parallel.

Pp. 55, 62 and 94: For "malazeit" read monazite.

P. 85, last line, also in the Bibliography: For "Klément and Regnard" read Klément et Renard.

Additions: P. 26, § 72. The Japanese paper for cleaning lenses was named by the author, "Lens-Paper," Feb. 27, 1892, and may be properly called by that name wherever mentioned in the book.

P. 21, § 59. In § 59 reference is only made to the swaying of the image due to oblique light. It should be added that if the coarse or fine adjustment is imperfect, the object may sway even when the light is axial. Swaying with central light would serve to indicate the defective mechanism.

In Chapter IV, special attention should be called to the work of MacMunn on the Spectroscope in Medicine, and the Bibliography of works relating to the spectroscope given in it.

Add to the list of books on pp. 86–89:

Carpenter–Dallinger—The Microscope and its Revelations, by the late William B. Carpenter. Seventh edition, in which the first seven chapters have been entirely re-written, and the text throughout reconstructed, enlarged and revised by the Rev. W. H. Dallinger. London and Philadelphia, 1891.

This work deals very satisfactorily with the higher problems relating to the microscope, and is invaluable as a work of reference.

Griffith and Henfrey—The Micrographic Dictionary; a guide to the examination and investigation of the structure and nature of microscopic objects. Fourth edition, by Griffith, assisted by Berkeley and Jones. London, 1883.

Pelletan, J.—Manuel d'Histologie normale. Paris, 1878.

Situation of the Plates:

Plates I and II, opposite page 1.
Plate III, ... p. 29.
Plate IV, ... p. 36.
Plate V, ... p. 66.
Plate VI, ... p. 54.

THE MICROSCOPE AND HISTOLOGY.

CONTENTS OF PART I.

CHAPTER I.

		PAGE.
§§ 1–74.	The Microscope and its Parts—Care and Use, ...	1–28

CHAPTER II.

§§ 75–96. Interpretation of Appearances, ... 29–35

CHAPTER III.

§§ 97–127. Magnification, Micrometry and Drawing, ... 36–53

CHAPTER IV.

§§ 128–157. The Micro-Spectroscope and Micro-Polariscope, ... 54–65.

CHAPTER V.

§§ 158–200. Slides and Cover-Glasses, Mounting, Labeling, Cataloging and Storing Microscopical Preparations; Experiments in Micro-Chemistry, ... 66–85

Bibliography, ... 86–90

Index, ... 91–96

LIST OF ILLUSTRATIONS.

All of the Figures, except when otherwise indicated, are original, and were drawn by Mrs. Gage.

PLATES.

PLATE I.

Fig.
1. Double convex lens showing the principal plane, the principal focus, and the focal distance.
2. Converging lens showing formation of a virtual image.
3. Converging lens showing formation of a real image.
4. Simple microscope with retinal image, and its projection as a virtual image.
5. Compound microscope, tracing the rays from the object to the final, virtual image.
6. Huygenian ocular or eye-piece, showing action of field-lens (Ross).
7. Huygenian ocular showing the eye-point.

PLATE II.

9. Tripod magnifier.
10. Stand of a compound microscope with names of parts.
11. Section of stage of compound microscope showing proper position of diaphragms.
12. Section of a low, dry objective and reflected light.
13. Section of an adjustable, immersion objective, transmitted axial and oblique light.
14. Diagram showing how to put on a cover-glass.
15. Slides showing how to enclose the lines of a micrometer or of some part of a preparation by a small ring.
16. Double eye-shade.

PLATE III.

20–22. Sectional views of the Abbe illuminator showing various methods of illumination,—with parallel rays of central light, with oblique light, with converging rays, and for dark-ground illumination.
23. Letters mounted in stairs to show order of coming into focus.
24. Glass rod in air and in glycerin.
25. Glass rod coated with collodion to show double contour.
26. Blood corpuscles on edge, to show surface and optical sections.
27. Wollaston's camera lucida in section, showing the overlapping fields.
28. Position of the microscope for determining magnification with Wollaston's camera lucida; also the necessity of a standard distance at which to measure the image.
29. Figures of the image of the stage and ocular micrometers, showing correct mutual arrangement of lines in determining the ocular micrometer valuation.

PLATE IV.

30. Sectional view of the Abbe camera lucida with a 45° mirror and a horizontal drawing surface.
31. Geometrical figure of the preceding showing the angles made by the axial ray with the drawing surface and with the mirror.
32. Sectional view of the Abbe camera lucida with a 35° mirror, showing the necessary elevation of the drawing surface to avoid distortion.
33. Geometrical figure of the preceding showing angles of axial ray and of drawing board, and that the drawing board must be raised twice as many degrees as the mirror is depressed below 45°.
34. Diagram showing arrangement of drawing board with mirror at 35° and with the microscope inclined 30° (Mrs. Gage).
35. Upper view of the prism of the camera lucida.
36. Eye-point of an ocular.
37. Quadrant with graduations to be added to the mirror of the Abbe camera lucida to determine the inclination of the mirror.

PLATE V.

41. Effect of the cover-glass on the rays from the object to the objective (Ross).
42. Direction of the rays from an object through a cover-glass in a dry objective.
43. Direction of the rays with a water immersion objective.
44. Direction of the rays with a homogeneous immersion objective. (Fig. 42-44 are modified from Ellenberger).
45. Absorption spectrum of arterial and venous blood; some of the principal Fraunhofer lines and an Ångström scale are also shown. (From Gamgee and MacMunn).
46. Centering card.
47. Small spirit lamp used as a reagent bottle for Canada balsam, glycerin jelly, shellac cement, etc.
48. Pipette or dropper for delivering small quantities of any liquid.
49. Slide and cover-glass showing the method of irrigation.
50. Showing the method of anchoring the cover-glass previous to sealing glycerin-mounted objects.
51-56. Various apparatus for the study of fibrin and the counting of blood corpuscles. (These figures appertain to Part II).

FIGURES IN THE TEXT.

FIG.		PAGE.
8.	Triplet for the pocket (Bausch & Lomb Optical Company),	2
8a.	Simple microscope with stand (R. & J. Beck),	2
8b.	Figure showing parts included in tube-length by various opticians,	6
17.	Double nose-piece or revolver (Bausch & Lomb Optical Co.),	11
18.	Ward's eye-shade (Bausch & Lomb Optical Co.),	27
19.	Oil-globule and air-bubble, with oblique light,	32
38.	Cover-glass measurer (Edward Bausch),	69
38a.	Turn-table for sealing cover-glass, etc., (James W. Queen & Co.),	71
39.	Cabinet for specimens,	80
40.	Cabinet drawer, face and sectional view,	81
57.	Arranging and labeling serial sections,	78

THE MICROSCOPE AND HISTOLOGY.

CHAPTER I.

THE MICROSCOPE AND ITS PARTS—CARE AND USE.

APPARATUS AND MATERIAL FOR THIS CHAPTER.

A simple microscope (§ 2, 4); A compound microscope with nose-piece (Fig. 17), eye-shade (Fig. 16, 18), achromatic (§ 12), apochromatic (§ 14), dry (§ 9), immersion (§10), unadjustable and adjustable objectives (§ 15, 16), Huygenian or negative (§ 20, 22), positive (§ 21) and compensation oculars (§ 23), Abbe illuminator (54), homogeneous immersion liquid (§ 10, 65–69), benzine and distilled water (§ 64, 69). Mounted letters or figures (§ 34); Ground-glass and Japanese filter or bibulous paper (§ 34, 72); Mounted preparation of fly's wing (§ 50); Mounted preparation of *Pleurasigma* (§ 52, 53, 58); Stage or ocular micrometer with lines filled with graphite (§ 52, 53, 59); Glass slides and cover glasses (§ 52); 10 per ct. solution of salicylic acid in 95 per ct. alcohol (§ 60); Preparation of stained microbes (§ 67); Vial of equal parts olive or cotton seed oil and benzine (§ 71).

Of the above, the laboratory furnishes all except the tripod magnifier, the glass slides and cover-glasses; these must be obtained by the student.

A MICROSCOPE.

§ 1. A Microscope is an optical apparatus with which one may obtain a clear image of a near object, the image being always larger than the object; that is, it enables the eye to see an object under a greatly increased visual angle, as if the object were brought very close to the eye without affecting the distinctness of vision. Whenever the microscope is used for observation, the eye of the observer forms an integral part of the optical combination (Pl. I, Fig. 4 and 5).

§ 2. A Simple Microscope.—With this an enlarged, erect image of an object may be seen. It always consists of one or more converging lenses or lens-systems (Pl. I, Fig. 1, 2 and 4), and the object must be placed within the principal focus (§ 4). The simple microscope may be held in the hand or it may be mounted in some way to facilitate its use (Fig. 8ᵃ).

§ 3. A Compound Microscope.—This enables one to see an enlarged, inverted image. It always consists of two optical parts,—an *objective*, to produce an enlarged, inverted, real image of the object, and an *ocular* acting in general like a simple microscope to magnify this real image (Pl. I, Fig. 5). There is also usually present a mirror, or both a mirror and some form of condenser or illuminator for lighting the object. The stand of the microscope consists of certain mechanical arrangements for holding the optical parts and for the more satisfactory use of them (Pl. II, Fig. 10).

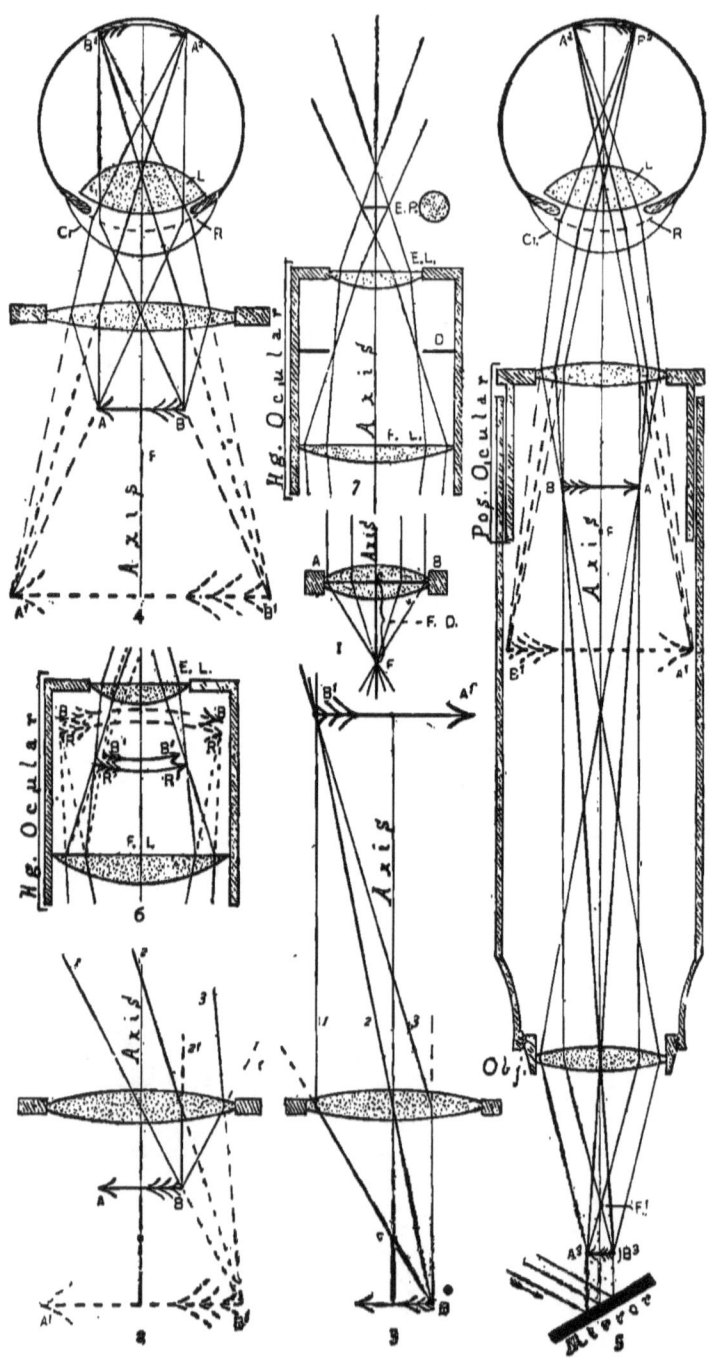

EXPLANATION OF PLATE I.

In all of the figures, Virtual Images and rays traced backward or produced rays, are indicated by dotted or broken lines, real rays or paths of rays by full or unbroken lines.

Fig. 1. Sectional view of a double convex lens showing: (A B) the *principal plane* at which the refractions of the curved surfaces are most conveniently shown; (c) Optical Center of the lens. Rays traversing this center undergo no deviation.

Axis. Principal optic axis of the lens, *i. e.*, line connecting the centers of curvature of the two surfaces of the lens. The axis traverses the optical center and the principal focal point or focus (F). F. Principal focal point or focus, *i. e.*, the point where central parallel rays are brought to a focus.

F D. Principal focal distance, or focal length, *i. e.*, the distance between the center of the lens (c) and the principal focus (F).

Fig. 2. Convex lens showing the position of the object (A B) within the principal focus and the course of rays in the formation of a virtual image.

A B. The object placed between the lens and its focus; A′ B′ virtual image formed by tracing the rays backward. It appears on the same side of the lens as the object, and is erect (§ 4).

Axis. The optic axis of the lens. The principal focus is represented by a dot on the axis between the object and virtual image.

1, 2, 3. Rays from the point B of the object. They are diverging after traversing the lens, but not so divergent as if no lens were present, as is shown by the dotted lines. Ray 1 traverses the center of the lens, and is therefore not deviated.

Fig. 3. Convex lens showing the position of the object (A-B) outside the principal focus (F), and the course of the rays in the formation of real images. To avoid confusion the rays are drawn from only one point, as in Fig. 2.

A B. Object outside the principal focus. B′ A′. Real, enlarged image on the opposite side of the lens.

Axis. Principal optic axis. 1, 2, 3. Rays after traversing the lens. They are converging, and consequently form a real image. The dotted lines and the line 2 give the direction of the rays unaffected by the lens. F. The principal focus.

Fig. 4. Diagram of the simple microscope showing the course of the rays and all the images, and that the eye forms an integral part of it.

A B. The object within the principal focus. A′ B′. The virtual image on the same side of the lens as the object. It is indicated with dotted lines, as it has no actual existence.

$A^2 B^2$. Retinal image of the object (A B). The virtual image is simply a projection of the retinal image in the field of vision.

Axis. The principal optic axis of the microscope and of the eye. Cr. Cornea of the eye. L. Crystalline lens of the eye. R. Ideal refracting surface at which all the refractions of the eye may be assumed to take place.

Fig. 5. Diagram of a compound microscope, showing the course of the rays from the object ($A^3 B^1$) through the objective to the real image (B A), thence through the ocular and into the eye to the retinal image ($A^2 B^2$), and the projection of the retinal image into the field of vision as the virtual image (B′ A′).

$A^3 B^3$. The object. $A^2 B^2$. The retinal image of the inverted real image, B A, formed by the objective. B′ A′. The inverted virtual image, a projection of the retinal image.

Axis. The optic axis of the microscope and the eye.

Cr. Cornea of the eye. L. Crystalline lens of the eye. R. Single, ideal, refracting surface at which all the refractions of the eye may be supposed to take place.

F. The principal focus of the positive ocular. F′. The principal focus of the objective.

Mirror. The mirror reflecting parallel rays to the object. The light is central (§ 42).

Pos. Ocular. An ocular in which the real image is formed outside the ocular. Compare the positive ocular with the simple microscope (Fig. 4).

Fig. 6. *Hg. Ocular.* Huygenian ocular showing the general character of a negative ocular, and the action of the field and eye-lenses. (From Carpenter, after A. Ross).

B B. Blue image, convex to the eye-lens, that would be formed if no field-lens were present.

R R. Red image, convex to the eye-lens, that would be formed but for the presence of the field-lens. B B and R R show also that the objective is over-corrected for the blue rays, as the blue image is formed farther from the objective than the red image. As blue rays are more refrangible than red, the image would naturally be nearer the objective than the red image.

B′ B′ R′ R′. Blue and red real images as actually formed under the influence of the field-lens. Both are concave to the eye-lens, and "as the focus of the eye-lens is shorter for blue rays than for red rays by just the amount of the difference in the place of these images, their rays, after refraction by it, enter the eye in a parallel direction, and produce a picture free from false color." The field-lens also aids in rendering the field flat.

E L. Eye-lens. F L. Field-lens.

Fig. 7. Sectional view of a Huygenian ocular (*Hg. ocular*), to show the formation of the Eye-Point.

Axis. Optic axis of the ocular. D. Diaphragm of the ocular. E L. Eye-lens. F L. Field-lens.

E P. Eye-point. As seen in section, it appears something like an hour-glass. When seen as in looking at the ocular, *i. e.*, in transection, it appears as a circle of light. It is at the point where most rays cross.

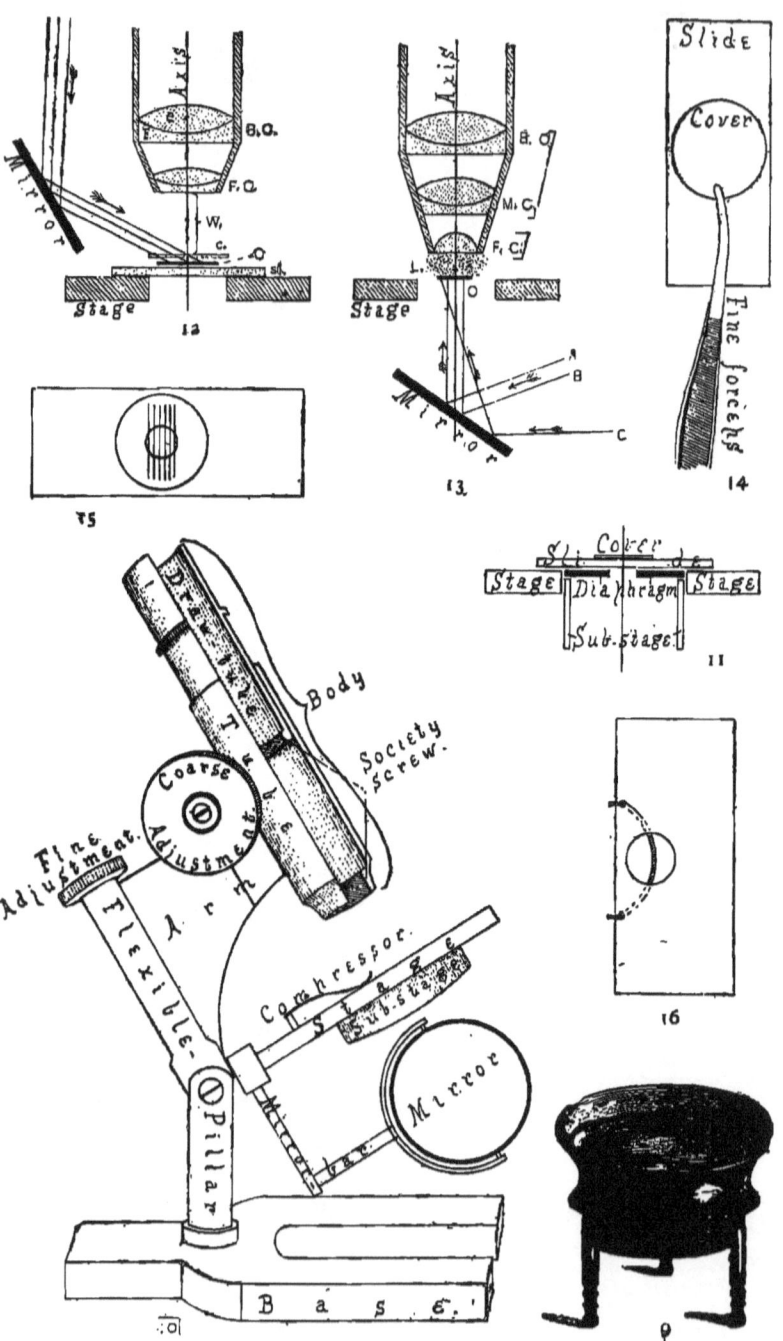

EXPLANATION OF PLATE II.

Fig. 9. *Tripod Magnifier*. The frame holding the lenses may be raised or lowered (focused) by screwing it up or down in the outside ring.

Fig. 10. *Stand*. That is the mechaical parts of a simple form of compound microscope, with the names of the parts written upon them.

Arm. The part connecting the body or tube to the pillar.

Base. The part of the stand on which it rests. It should be heavy and so formed that it will give steadiness, and not be in the way of the mirror.

Body. The tube and draw-tube together. Also frequently called the *tube*.

Coarse Adjustment. The rack and pinion for moving the tube or body rapidly up or down.

Fine Adjustment. The micrometer screw arrangement for moving the body or tube of the microscope slowly up or down.

Compressor. One of the pair of light springs to hold the preparation in position on the stage.

Flexible Pillar. The pillar of the microscope, with a joint to incline the microscope.

Mirror. The movable mirror with plane and concave face for lighting the object.

Mirror Bar. The bar supporting the mirror.

Society Screw. The screws at the lower end of the draw-tube and the main tube or body tube for receiving the objective.

Stage. The horizontal plate for supporting the object.

Substage. The cylinder below the stage for diaphragms, the illuminator and other substage accessories.

Fig. 11. Sectional view of the stage to show the relative position of the preparation and the diaphragms necessary to insure the most satisfactory lighting when a mirror is used.

Fig. 12. Section of a dry objective showing working distance and lighting by reflected light.

Axis. The optic axis of the objective.

B C. *Back Combination*, composed of a plano-concave of flint glass (F), and a double convex of crown glass (c).

F C. *Front Combination*.

C O sl. The cover-glass, object and slide.

Mirror. The mirror is represented as above the stage, and as reflecting parallel rays from its plane face upon the object.

Stage. Section of the stage of the microscope.

W. The *Working Distance*, that is the distance from the front of the objective to the object when the objective is in focus (¿ 38).

Fig. 13. Sectional view of an *Immersion Adjustable Objective*, and the object lighted with axial or central and with oblique light.

Axis. The optic axis of the objective.

B C, M C, F C. The back, middle and front combinations of the objective. In this case the front is not a combination, but a single plano-convex lens.

A B. Parallel rays reflected by the mirror axially or centrally upon the object.

C. Ray reflected to the object obliquely.

I. Immersion fluid between the front of the objective and the cover-glass or object O).

Mirror. The mirror of the microscope.

O. *Object*. It is represented without a cover-glass. Ordinarily objects are covered whether examined with immersion or with dry objectives.

Stage. Section of the stage of the microscope.

Fig. 14. Diagram showing how to place a cover-glass upon an object with fine forceps.

Fig. 15. Diagram showing how to enclose the lines of a micrometer, or of some part of a preparation by a small ring to facilitate finding it under the microscope (¿ 32).

Fig. 16. Double eye-shade (¿ 73). This is made by cutting a hole slightly larger than the tube near one edge. A rubber band is then used to loop around the tube and holding the screen from falling over in front. It is desirable to have the screen covered with velveteen.

SIMPLE MICROSCOPE: EXPERIMENTS.

§ 4. Employ a tripod or other simple microscope, and for object a printed page. Hold the eye about two centimeters from the upper surface of the magnifier, then alternately raise and lower the magnifier until a clear image may be seen. (This mutual arrangement of microscope and object so that a clear image may be seen, is called focusing, see § 37). When a clear image is seen, note that the letters appear as with the unaided eye except that they are larger, and the letters appear erect or right side up, instead of being inverted, as with the compound microscope (§ 3, 34).

Hold the simple microscope directly toward the sun and move it away from and toward a piece of printed paper until the smallest bright point on the paper is obtained. This is the *burning point* or *focus*, and as the rays of the sun are nearly parallel, the burning point represents approximately the principal focus (Fig. 1). Without changing the position of the paper or the magnifier, look into the magnifier and note that the letters are very indistinct or invisible. Move the magnifier a centimeter or two farther from the paper and no image can be seen. Now move the magnifier closer to the paper, that is, so that it is less than the focal distance from the paper, and the letters will appear distinct. This shows that in order to see a distinct image with a simple microscope, the object must always be nearer to it than its principal focal point. Or, in other words, the object must be within the principal focus. Compare § 34.

FIG. 8. — *Achromatic Triplet for the Pocket.*

After getting as clear an image as possible with a simple microscope, do not change the position of the microscope but move the eye nearer and farther from it, and note that when the eye is in one position, the largest field may be seen (§ 33). This position corresponds to the eye-point (§ 36) of an ocular, and is the point at which the largest number of rays from the microscope enter the eye. Note that the image appears on the same side of the magnifier as the object (§ 34).

Simple microscopes are very convenient when only a small magnification (Ch. III) is desired, as for dissecting. Achromatic triplets are excellent and convenient for the pocket (Fig. 8). For use in conjunction with a compound microscope, the tripod

FIG. 8a.—*Simple Microscope with Special Mechanical Mounting to Hold and Focus the Magnifier and to Support and Light the Object.*

magnifier (Plate II, Fig. 9) is one of the best forms. For many purposes a special mechanical mounting like that of Fig. 8ª is to be preferred.

COMPOUND MICROSCOPE.

MECHANICAL PARTS.

§ 5. **The Mechanical Parts** of a laboratory, compound microscope are shown in Pl. II, Fig. 10, and are described in the explanation of that figure. The student should study the figure with a microscope before him and become thoroughly familiar with the names of all the parts.

OPTICAL PARTS.

§ 6. **Microscopic Objectives.**—These consist of a converging lens or of one or more converging lens-systems, which give an enlarged, inverted, real image of the object (Pl. I, Fig. 3 and 5). And as for the formation of real images generally, the object must be placed outside the principal focus, instead of within it, as for the simple microscope. (See §§ 4, 34).

Modern microscopic objectives usually consist of two or more systems or combinations of lenses, the one next the object being called the *front combination* or lens, the one farthest from the object and nearest the ocular, the *back combination* or system. There may be also one or more intermediate systems. Each combination is, in general, composed of a convex and a concave lens. The combined action of the systems serves to produce an image free from color and from spherical distortion. In the ordinary achromatic objectives the convex lenses are of crown and the concave lenses of flint glass (Pl. II, Fig. 12, 13).

NOMENCLATURE OR TERMINOLOGY OF OBJECTIVES.

§ 7. **Equivalent Focus.**—In America, England, and sometimes also on the Continent, objectives are designated by their equivalent focal length. This length is given either in inches (usually contracted to in.) or in millimeters (mm.). Thus: An objective designated $\frac{1}{12}$ in. or 2 mm., indicates that the objective produces a real image of the same size as is produced by a simple converging lens whose principal focal distance is $\frac{1}{12}$ inch or 2 millimeters (Pl. I, Fig. 1). An objective marked 3 in. or 75 mm., produces approximately the same sized real image as a simple converging lens of 3 inches or 75 millimeters focal length. And in accordance with the law that the relative size of object and image vary directly as their distance from the center of the lens (Pl. I, Fig. 2, 3), it follows that the less the focal distance of the simple lens or of the equivalent focal distance of the objective, the greater is the size of the real image.

§ 8. **Numbering or Lettering Objectives.**—Instead of designating objectives by their equivalent focus, many Continental opticians use letters or figures for this purpose. With this method the smaller the number, or the earlier in the alphabet the letter, the lower is the power of the objective. (See further in Ch. III, for the power or magnification of Objectives.) This method is entirely arbitrary and does not, like the one above, give direct information concerning the objective.

§ 9. **Dry Objectives.**—These are objectives in which the space between the front of the objective and the object or cover-glass is filled with air (Pl. II, Fig. 12). Most objectives of low and medium power (*i. e.*, ⅕th in. or 3 mm. and lower powers) are dry.

§ 10. **Immersion Objectives.**—An immersion objective is one with which there is some liquid placed between the front of the objective and the object or cover-glass. The most common immersion objectives are those (A) in which water is used as the immersion fluid, and (B) where some liquid is used having the same refractive and dispersive power as the front lens of the objective. Such a liquid is called homogeneous, as it is optically homogeneous with the front glass of the objective. It may consist of thickened cedar-wood oil or of glycerin containing some salt, as stannous chloride, in solution. When oil is used as the immersion fluid the objectives are frequently called oil-immersion objectives. The disturbing effect of the cover-glass (§§ 16, 63) is almost wholly eliminated by the use of homogeneous immersion objectives.

The course of the rays of light from the object to the objective with dry and different forms of immersion objectives is shown in Pl. V, Fig. 42, 43, 44.

§ 11. **Non-Achromatic Objectives.**—These are objectives in which the chromatic aberration is not corrected, and the image produced is bordered by colored fringes. They show also spherical aberration and are used only on very cheap microscopes.

§ 12. **Achromatic Objectives.**—In these the chromatic and the spherical aberration are both largely eliminated by combining concave and convex lenses of different kinds of glass "so disposed that their opposite aberrations shall correct each other." All the better forms of objectives are achromatic and also aplanatic (§ 13).

§ 13. **Aplanatic Objectives, etc.**—These are objectives or other pieces of optical apparatus (oculars, illuminators, etc.), in which the spherical distortion is wholly or nearly eliminated, as in § 12. Such pieces of apparatus are usually achromatic also.

§ 14. **Apochromatic Objectives.**—A term used by Abbe to designate a new form of objectives made by combining new kinds of glass with a natural mineral (Calcium fluoride, Fluorite, or Fluor-spar). The name, Apochromatic, is used to indicate the higher kind of achromatism in which rays of three spectral colors are combined at one focus, instead of rays of two colors, as in the ordinary achromatic objectives.

The special characteristics of these objectives, when used with the "compensating oculars" (§ 23), are as follows:

(1) *Three* rays of different color are brought to one focus, leaving a small tertiary spectrum only, while with objectives as formerly made from crown and flint glass, only *two* different colors could be brought to the same focus.

(2) In these objectives the correction of the spherical aberration is obtained for *two* different colors in the brightest part of the spectrum, and the objective shows the same degree of chromatic correction for the marginal as for the central part of the aperture. In the old objectives, correction of the spherical aberration was confined to rays of *one* color, the correction being made for the central part of the spectrum, the objective remaining *under*-corrected spherically for the red rays and *over*-corrected for the blue rays.

(3) The optical and chemical foci are identical, and the image formed by the chemical rays is much more perfect than with the old objectives, hence the new objectives are well adapted to photography.

(4) These objectives admit of the use of very high oculars, and seem to be a considerable improvement over those made in the old way with crown and flint glass. According to Dippel (Z. w. M. 1886, p. 300), dry apochromatic objectives give as clear images as the same power water immersion objectives of the old form.

§ 15. **Non-Adjustable or Unadjustable Objectives.**—Objectives in which the lenses or lens systems are permanently fixed in their mounting so that their relative position always remains the same. Low power objectives and those with homogeneous immersion are mostly non-adjustable. For beginners and those unskilled in manipulating adjustable (§ 16) objectives, non-adjustable ones are more satisfactory, as the optician has put the lenses in such a position that the most satisfactory results may be obtained when the proper thickness of cover-glass and tube-length are employed. (See § 17 and table and figure of tube-length and thickness of cover-glass below.)

§ 16. **Adjustable Objectives.**—An adjustable objective is one in which the distance between the systems of lenses (usually the front and the back systems) may be changed by the observer at pleasure. The object of this adjustment is to correct or compensate for the displacement of the rays of light produced by the mounting medium and the cover-glass after the rays have left the object. It is also to compensate for variations in "tube length." See § 17. As the displacement of the rays by the cover-glass is the most constant and important, these objectives are usually designated as having cover-glass adjustment or correction. (Pl. II, Fig. 13. See also practical work, § 63.)

§ 17. **Tube-Length and Thickness of Cover-Glasses.**—"In the construction of microscopic objectives, the corrections must be made for the formation of the image at a definite distance, or in other words the tube of the microscope on which the objective is to be used must have a definite length. Consequently the microscopist must know and use this distance or 'microscopical tube-length' to obtain the best results in using any objective in practical work." Unfortunately different opticians have selected different tube-lengths and also different points between which the distance is measured, so that one must know what is meant by the tube-length of each optician whose objectives are used. See table.

The thickness of cover-glass used on an object, (see Ch. V, on mounting), except with homogeneous immersion objectives, has a marked effect on the light passing from the object (Pl. V, Fig. 41). To compensate for this the relative positions of the systems composing the objective are different from what they would be if the object were uncovered. Consequently, in non-adjustable objectives some standard thickness of cover-glass is chosen by each optician and the position of the systems arranged accordingly. With such an objective the image of an uncovered object would be less distinct than a covered one, and the same result would follow the use of a cover-glass much too thick (§ 63, Fig. 41).

Length in Millimeters and Parts Included in "Tube-Length" by Various Opticians.*

FIG. 8b.

	Pts. included in "Tube-length." See Diagram.	"Tube-length" in Millimeters.
a-d	Grunow, New York	203 mm.
	Nachet et Fils, Paris	146 or 200 mm.
	Powell and Lealand, London	254 mm.
	C. Reichert, Vienna	160 to 180 mm.
	W. Wales, New York	254 mm.
b-d	Bausch & Lomb Opt. Co., Rochester .	216 mm.
	Bézu, Hausser et Cie, Paris†	220 mm.
	Klönne und Müller, Berlin . . .	160-180 or 254 mm.
	W. & H. Seibert, Wetzlar	190 mm.
	Swift & Son, London	165 to 228½ mm.
	C. Zeiss, Jena	160 or 250 mm.
a-g	Gundlach Optical Co., Rochester . .	254 mm.
a-g	R. Winkel, Göttingen	220 mm.
c-d	Ross & Co., London	254 mm.
c-e	R. & J. Beck, London	254 mm.
c-g	H. R. Spencer & Co., Geneva, N. Y.	254 mm.
c-f	J. Green, Brooklyn ‡	254 mm.
c'-e	E. Leitz, Wetzlar	125-180 mm.
	For oil immersions	160 mm.

Thickness of Cover Glass for which Non-Adjustable Objectives are Corrected by Various Opticians.

$\frac{18}{100}$ mm.	J. Green, Brooklyn.
	J. Grunow, New York.
	Powell and Lealand, London.
	H. R. Spencer & Co., Geneva, N. Y.
	W. Wales, New York.
$\frac{18}{100}$ mm.	Klönne und Müller, Berlin.
$\frac{17}{100}$ mm.	E. Leitz, Wetzlar (when tube 160-170 mm.).
$\frac{17}{100}$ mm.	R. Winkel, Göttingen, Germany.
$\frac{15-21}{100}$ mm.	Ross & Co., London.
$\frac{16}{100}$ mm.	Bausch & Lomb Optical Co., Rochester.
$\frac{12-20}{100}$ mm.	C. Zeiss, Jena ($\frac{10}{100}$ mm. for apochromatic oil immersions).
$\frac{16-18}{100}$ mm.	C. Reichert, Vienna.
$\frac{15}{100}$ mm.	Gundlach Optical Co., Rochester.
	W. & H. Siebert, Wetzlar.
	R. & J. Beck, London.
$\frac{12-17}{100}$ mm.	J. Zentmayer, Philadelphia.
$\frac{10-12½}{100}$ mm.	Nachet et Fils, Paris.
	Bézu, Hausser et Cie, Paris.
$\frac{10}{100}$ mm.	Swift & Son, London.

§ 18. **Aperture of Objectives.**—The angular aperture or angle of aperture of an objective is the angle "contained, in each case, between the most diverging of the rays issuing from the axial point of an object [*i.e.*, a point in the object situated on the extended optic axis of the microscope], that can enter the objective and take part in the formation of an image." (C.)

---

* The information contained in these tables was very kindly furnished by the opticians named.
† Successors to Hartnack.
‡ Successor to Tolles.

According to some other authors the angle of aperture is the angle between the extreme rays from the focal point which can be transmitted through the entire objective. This would give a somewhat greater angle than by the first method as the focal point of the objective is nearer to it than the axial point of the object (Pl. I, Fig. 1, 3 and 5).

In general, the angle increases with the size of the lenses forming the objective and the shortness of the equivalent focal distance (§ 7). If all objectives were dry or all water or homogeneous immersion a comparison of the angular aperture would give one a good idea of the relative number of image forming rays transmitted by different objectives; but as some are dry, others water and still others homogeneous immersion, one can see at a glance (see Pl. V, Fig. 42, 43, 44) that other things being equal, the dry objective (Fig. 42) receives less light than the water immersion, and the water immersion (Fig. 43) less than the homogeneous immersion (Fig. 44). In order to render comparison accurate between different kinds of objectives, Professor Abbe takes into consideration the rays actually passing from the back combination of the objective to form the real image; he thus takes into account the medium in front of the objective as well as the angular aperture. The term "*numerical aperture*" was introduced by Abbe to indicate the capacity of an optical instrument "for receiving rays from the object and transmitting them to the image, and the aperture of a microscopic objective is therefore determined by the ratio between its focal length and the diameter of the emergent pencil at the point of its emergence, that is the utilized diameter of a single-lens objective or of the back lens of a compound objective."

Numerical Aperture (abbreviated N.A.) is then the ratio of the diameter of the emergent pencil to the focal length of the lens, or as usually expressed, the factors being more readily obtainable, it is the index of refraction of the medium in front of the objective (*i.e.*, air for dry, and water or homogeneous fluid for immersion objectives) multiplied by the sine of half the angle of aperture. The usual formula is N.A. $= n$ sin u; N.A. representing numerical aperture, n the index of refraction of the substance in front of the objective, and u the semi-angle of aperture.

For example, take three objectives each of 3 mm. equivalent focus, one being a dry, one a water immersion, and one a homogeneous immersion. Suppose that the dry objective has an angular aperture of 106°, the water immersion of 94° and the homogeneous immersion of 90°. Simply compared as to their angular aperture, without regard to the medium in front of the objective, it would look as if the dry objective would actually take in and transmit a wider pencil of light than either of the others. However, if the medium in front of the objective is considered, that is to say, if the numerical instead of the angular apertures are compared, the results would be as follows; Numerical Aperture of a dry objective of 106°, N.A. $= n$ sin u. In the case of dry objectives the medium in front of the objective being air the index of refraction is unity, whence $n=1$. Half the angular aperture is $\frac{106°}{2} = 53°$. By consulting a table of natural sines it will be found that the sine of 53° is 0.799, whence N.A. $= \sin n$ or $1 \times \sin u$ or $0.799 = 0.799$.

With the water immersion objective in the same way N.A. $= n$ sin u. In this case the medium in front of the objective is water, and its index of refraction is 1.33, whence $n = 1.33$. Half the angular aperture is $\frac{94°}{2} = 47°$, and by consulting a table of natural sines, the sine of 47° is found to be 0.731 *i.e.* sin $u = 0.731$, whence N.A. $= n$ or 1.33 $\times$ sin u or $0.731 = 0.972$.

With the oil immersion in the same way N.A. $= n$ sin u; n or the index of refraction of the homogeneous fluid in front of the objective is 1.52, and the semi-angle

of aperture is $\frac{90°}{2} = 45°$. The sine of 45° is 0.707 whence N.A. $= n$ or $1.52 \times \sin u$ or $0.707 = 1.074$.

By comparing these numerical apertures: Dry 0.799, water 0.972, homogeneous immersion 1.074 the same idea of the real light efficiency and image power of the different objectives is obtained, as in the graphic representations shown in Pl. V., Fig. 42, 43, 44.

THE OCULAR.

§ 19. A **Microscopic Ocular** or **Eye-Piece** consists of one or more converging lenses or lens systems, the combined action of which is, like that of a simple microscope, to magnify the real image formed by the objective.

Depending upon the relation and action of the different lenses forming oculars, they are divided into two great groups, *negative* and *positive*.

§ 20. **Negative Oculars**, are those in which the real, inverted image is formed within the ocular, the lower or field-lens serving to collect the image-forming rays somewhat so that the real image is smaller than as if the field-lens were absent (Pl. I, Fig. 6). As the field-lens of the ocular aids in the formation of the real image it is considered by some to form a part of the objective rather than of the ocular. The upper or eye-lens of the ocular magnifies the real image.

§ 21. **Positive Oculars** are those in which the real, inverted image of the objective is formed outside the ocular, and the entire system of ocular lenses magnifies the real image like a simple microscope (Pl. I, Fig. 5).

Positive and negative oculars may be readily distinguished, as a positive ocular may be used as a simple microscope, while a negative ocular cannot be so used when its field glass is in the natural position toward the object. By turning the eye-lens toward the object and looking into the field-lens an image may be seen, however.

Special names have also been applied to oculars, depending upon the designer, the construction, or the special use to which the ocular is to be applied. The following are used in the anatomical department of Cornell University :—*

---

*In works and catalogues concerning the microscope and microscopic apparatus, and in articles upon the microscope in periodicals, various forms of oculars or eye-pieces are so frequently mentioned, without explanation or definition, that it seemed worth while to give a list, with the French and German equivalents, and a brief statement of their character.

Achromatic Ocular; Fr. oculaire achromatique; Ger. achromatisches Okular. Oculars in which chromatic aberration is wholly or nearly eliminated. *Aplanatic* Ocular; Fr. Oculaire aplanatique; Ger. aplanatisches Okular (see § 13). *Binocular, stereoscopic* Ocular; Fr. Oculaire binoculaire stereoscopique; Ger. stereoskopisches Doppel-Okular. An ocular consisting of two oculars about as far apart as the two eyes. These are connected with a single tube which fits a monocular microscope. By an arrangement of prisms the image forming rays are divided, half being sent to each eye. The most satisfactory form was worked out by Tolles and is constructed on true stereotomic principles, both fields being equally illuminated. His ocular is also erecting. *Campani's* Ocular (See Huygenian Ocular). . *Compound* Ocular; Fr. Oculaire composé; Ger. zusammengesetztes Okular. An ocular of two or more lenses, *e. g.*, the Huygenian (see Fig. 5 and 6). *Deep* Ocular, see high ocular. *Erecting* Ocular; Fr. Oculaire redresseur; Ger. bildumkehrendes Okular. An ocular with which an erecting prism is connected

§ 22. **Huygenian Ocular.**—A negative ocular designed by Huygens for the telescope, but adapted also to the microscope. It is the one now most commonly employed. It consists of a field-lens or collective (Pl. I, Fig. 6), aiding the objective in forming the real image, and an eye-lens which magnifies the real image. While the field-lens aids the objective in the formation of the real, inverted image, and increases the field of view; it also combines with the eye-lens in rendering the image achromatic (§ 35).

§ 23. **Compensating Oculars.**—These are oculars specially constructed for use with the apochromatic objectives. They compensate for aberrations outside the axis which could not be so readily eliminated in the objective itself. Oculars of this kind, magnifying but once or twice, are made for use with high powers, for the sake of the large field in finding objects; they are called *searching oculars*; those ordinarily used for observation are in contradistinction called *working oculars*. Part of the compensating oculars are positive and part negative.

§ 24. **Projection Oculars.**—These are oculars especially designed for projecting a microscopic image on the screen for class demonstrations, or for photographing so that the image is erect as with the simple microscope. Such oculars are most common on dissecting microscopes. *Goniometer* Ocular; Fr. Oculaire à goniomètre; Ger. Goniometer-Okular. An ocular with goniometer for measuring the angles of minute crystals. *High* Ocular, sometimes called a deep ocular. One that magnifies the real image considerably, *i. e.*, 10 to 20 fold. *Huygenian* Ocular, Huygens' O., Campani's O.; Fr. Oculaire d'Huygens, o. de Campani; Ger. Huygens'sches Okular, Campaniches Okular, see § 22. *Kellner's* Ocular, see orthoscopic ocular. *Low* Ocular, also called shallow ocular. An ocular which magnifies the real image only moderately, *i. e.*, 2 to 8 fold. *Micrometer* or *micrometric* Ocular; Fr. Oculaire micrometrique or à micromètre; Ger. Mikrometer-Okular, see § 25. *Microscopic* Ocular; Fr. Oculaire microscopique; Ger. Mikroskopisches Okular. An ocular for the microscope instead of one for a telescope. *Negative* Ocular, see § 21. *Orthoscopic* Oculars; also called Kellner's Ocular; Fr. Oculaire orthoscopique; Ger. Kellner'sches oder Orthoskopisches Okular. An ocular with an eye-lens like one of the combinations of an objective (Pl. II, Fig. 12, 13) and a double convex field-lens. The field-lens is in the focus of the eye-lens and there is no diaphragm present. The field is large and flat. *Periscopic* Ocular; Fr. Oculaire periscopique; Ger. Periskopisches Okular. A positive ocular devised by Gundlach. It consists of a double convex field-lens and a triplet eye-lens. It gives a large flat field. *Positive* Ocular, see § 21. *Projection* Ocular; Ger. *Projections*-Okular, see § 24. *Ramsden's* Ocular; Fr. Oculaire de Ramsden; Ger. Ramsden'sches Okular. A positive ocular devised by Ramsden. It consists of two plano-convex lenses placed close together with the convex surfaces facing each other. Only the central part of the field is clear. *Searching* Ocular; Ger. Sucher-Okular, see § 23. *Shallow* Ocular, see low ocular. *Solid* Ocular, *holosteric* O.; Fr. Oculaire holostère; Ger. Holosterisches Okular, Vollglass-Okular. A negative eye-piece devised by Tolles. It consists of a solid piece of glass with a moderate curvature at one end for a field-lens, and the other end with a much greater curvature for an eye-lens. For a diaphragm, a groove is cut at the proper level and filled with black pigment. It is especially excellent where a high ocular is desired. *Spectral* or *spectroscopic* Ocular; Fr. Oculaire spectroscopique; Ger. Spectral-Okular, see Microspectroscope, Ch. IV. *Working* Ocular; Ger. Arbeits-Okular, see § 23.

with the microscope. While they are specially adapted for use with apochromatic objectives, they may also be used with ordinary achromatic objectives of large numerical aperture.

§ 25. **Micrometer Ocular.**—This is an ocular connected with an ocular micrometer. The micrometer may be removable, or it may be permanently in connection with the ocular, and arranged with a spring and screw, by which it may be moved back and forth across the field. (See Ch. III, under Micrometry).

§ 26. **Spectral or Spectroscopic Ocular.**—(See Micro-Spectroscope, Ch. IV).

DESIGNATION OF OCULARS.

§ 27. **Equivalent Focus.**—As with objectives some opticians designate the oculars by their equivalent focus (§ 7). With this method the power of the ocular varies inversely with the focal length, *i. e.*, the less the equivalent focus the greater the power, and the greater the focal length the lower the power.

§ 28. **Numbering and Lettering.**—Oculars like objectives may be numbered or lettered arbitrarily. When so designated, the smaller the number, or the earlier the letter in the alphabet, the lower the power of the ocular.

§ 29. **Magnification or Combined Magnification and Equivalent Focus.**—The compensating oculars are marked both with their equivalent focus and the amount they magnify the real image. Thus, an occular marked x 4, 45 mm., indicates that the eqnivalent focus is 45 milimeters, and that the real image of the objective is multiplied four-fold by the ocular.

The projection oculars are designated simply by the amount they multiply the real image of the objective. Thus for the short or 160 mm. tube-length they are, x 2, x 4; and for the long, or 250 mm. tube, they are x 3 and x 6. That is, the final image on the screen or the ground glass of the photographic camera will be 2, 3, 4, or 6 times greater than it would be if no ocular were used.

COMPOUND MICROSCOPE.

EXPERIMENTS

§ 30. **Putting an Objective in Position and Removing it.**—Elevate the body of the microscope by means of the coarse adjustment (Fig. 10), so that there may be plenty of room between its lower end and the stage. Grasp the objective lightly near its lower end with two fingers of the left hand, and hold it against the nut in the lower end of the body (Fig. 10). With two fingers of the right hand take hold of milled ring near the back or upper end of the objective, and screw it into the body of the microscope. Reverse this operation for removing the objective. By following this method the danger of dropping the objective will be avoided.

§ 31. **Putting an Ocular in Position and Removing it.**—Elevate the body of the microscope with the coarse adjustment (Fig. 10), so that the objective will be 2 cm. or more from the object—grasp the ocular by the milled ring next the eye-lens (Fig. 5), and the coarse adjustment or the tube of the microscope and gently force the ocular into

position. In removing the ocular, reverse the operation. If the above precautions are not taken, and the oculars fit snugly, there is danger in inserting them of forcing the body of the microscope downward and the objective upon the object.

§ 32. **Putting an Object under the Microscope.**—This is so placing an object under the simple microscope, or on the stage of the compound microscope, that it will be in the field of view when the microscope is in focus (§ 33).

With low powers, it is not difficult to get an object under the microscope. The difficulty increases, however, with the power of the microscope and the smallness of the object. It is usually necessary to move the object in various directions while looking into the microscope, in order to get it into the field. Time is usually saved by getting the object in the center of the field with a low objective before putting the high objective in position. This is greatly facilitated by using a double nose-piece, or revolver.*

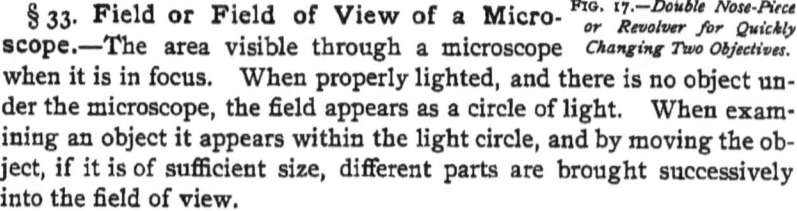

FIG. 17.—*Double Nose-Piece or Revolver for Quickly Changing Two Objectives.*

§ 33. **Field or Field of View of a Microscope.**—The area visible through a microscope when it is in focus. When properly lighted, and there is no object under the microscope, the field appears as a circle of light. When examining an object it appears within the light circle, and by moving the object, if it is of sufficient size, different parts are brought successively into the field of view.

In general, the greater the magnification of the entire microscope, whether the magnification is produced mainly by the objective, the ocular, or by increasing the tube-length, or by a combination of all three (see Ch. III, under magnification), the smaller is the field.

The size of the field is also dependent, in part, without regard to magnification, upon the size of the opening in the ocular diaphragm.

---

* As specimens are sometimes very small, or some part of a large specimen shows a particular structure with special excellence, it is desirable to mark the preparation so that the minute object or the part of a large object may be found quickly and with certainty. A simple way to do this is to find the object under the microscope, and then place a minute spot of black ink at one side. After this is done, remove the slide from the stage and surround the object with a ring of shellac cement, making the ring as small as possible and not cover the object. It will then always be known that the part to be examined is within the ring (B. 1, 47, C. 117). The enclosure in a ring may also be very elegantly done by the use of a marking apparatus like that of Winkel's (B. K. & S., p. 48), making use of either a diamond point or a delicate brush dipped in shellac or other cement.

Some oculars, as the orthoscopic and periscopic, are so constructed as to eliminate the ocular diaphragm, and in consequence, although this is not the sole cause, the field is considerably increased. The exact size of the field may be read off directly by putting a stage micrometer under the microscope and noting the number of spaces required to measure the diameter of the light circle.

The size of the field of the microscope as projected into the field of vision of the normal human eye (*i. e.*, the virtual image) may be determined by the use of the camera lucida with the drawing surface placed at the standard distance of 250 millimeters (Ch. III).

FUNCTION OF AN OBJECTIVE.

§ 34. Put a 2-in. (50 mm.) objective on the microscope, or screw off the front combination of a ¾-in. (18 mm.), and put the back-combination on the microscope for a low objective.

Place some printed letters or figures under the microscope, and light well. In place of an ocular, put a screen of ground glass, or a piece of Japanese or tissue paper, over the upper end of the body of the microscope.*

Lower the body by means of the coarse adjustment (Fig. 10), until the objective is within 2–3 cm. of the object on the stage. Look at the screen on the top of the body, holding the head about as far from it as for ordinary reading, and slowly elevate the body by means of the coarse adjustment until the image of the letters appears on the screen.

The image can be more clearly seen if the object is in a strong light and the screen in a moderate light, *i. e.*, if the top of the microscope is shaded.

The letters will appear as if printed on the ground glass or paper, but will be inverted (Fig. 5).

If the objective is not raised sufficiently, and the head is held too near the microscope, the objective will act as a simple microscope. If the letters are erect, and appear to be down in the microscope and not on the screen, hold the head farther from it, shade the latter, and raise the body of the microscope until the letters do appear on the screen.

To demonstrate that the object must be outside the principal focus with the compound microscope, remove the screen and turn the tube of the microscope directly toward the sun. Move the tube of the microscope with the coarse adjustment until the burning or focal point is

---

* Ground glass may be very easily prepared by placing some fine emery between two pieces of glass, wetting it with water and then rubbing the glasses together for a few minutes. If the glass becomes too opaque, it may be rendered more translucent by rubbing some oil upon it.

found (§ 4). Measure the distance from the paper object on the stage to the objective, and it will represent approximately the principal focal distance (Pl. I, Fig. 1). Replace the screen over the top of the tube, no image can be seen. Slowly raise the tube of the microscope and the image will finally appear. If the distance between the object and the objective is now taken, it will be found considerably greater than the principal focal distance (compare § 4).

Aerial Image.—After seeing the real image on the ground-glass, or paper, use the Japanese paper over about half of the opening of the tube of the microscope. Hold the eye about 250 mm. from the microscope as before and shade the top of the tube by holding the hand between it and the light, or in some other way. The real image can be seen in part as if on the paper and in part in the air. Move the paper so that the image of half a letter will be on the paper and half in the air. Another striking experiment is to have a small hole in the paper placed over the center of the tube opening, then if a printed word extends entirely across the diameter of the tube its central part may be seen in the air, the lateral parts on the paper. The advantage of the paper over part of the opening is to enable one to accommodate the eyes for the right distance. If the paper is absent the eyes adjust themselves for the light circle at the back of the objective, and the aerial image appears low in the tube. Furthermore, it is more difficult to see the aerial image in space than to see the image on the ground-glass or paper, for the eye must be held in the right position to receive the rays projected from the real image, while the granular surface of the glass and the delicate fibers of the paper reflect the rays irregularly, so that the image may be seen at almost any angle, as if the letters were actually printed on the paper or glass.

The function of an objective, as seen from these experiments, is to form an enlarged, inverted, real image of an object, this image being formed on the opposite side of the objective from the object (Fig. 5).

FUNCTION OF AN OCULAR.

§ 35. Using the same objective as for § 34, get as clear an image of the letters as possible on the Japanese paper screen. Look at the image with a simple microscope (Fig. 8 or 9) as if the image were an object. Observe that the image seen through the simple microscope is merely an enlargement of the one on the screen, and that the letters remain inverted, that is they appear as with the naked eye (§ 4). Remove the screen and observe the aerial image with the tripod.

Put an A, No. 1, 2 in. or 45 mm. ocular (*i. e.*, an ocular of low magnification) in position (§ 31). Hold the eye about 10 to 20 millimeters from the eye-lens and look into the microscope. The letters will ap-

pear as when the simple microscope was used (see above), the image will become more distinct by slightly raising the body of the microscope with the coarse adjustment.

The function of the Ocular, as seen from the above, is that of a simple microscope, viz.: It magnifies the real image formed by the objective as if that image were an object. Compare the image formed by the ocular (Fig. 5), and that formed by a simple microscope (Fig. 4).

It should be borne in mind, however, that the rays from an object as usually examined with a simple microscope, extend from the object in all directions, and no matter at what angle the simple microscope is held, provided it is sufficiently near and points toward the object, an image may be seen. The rays from a real image, however, are continued in certain definite lines and not in all directions; hence, in order to see the image with an ocular or simple microscope, or in order to see the aerial image with the unaided eye, the simple microscope, ocular or eye must be put in the path of the rays.

The field-lens of a Huygenian ocular makes the real image smaller and consequently increases the size of the field; it also makes the image brighter by contracting the area of the real image (Pl. I, Fig. 6). Demonstrate this by screwing off the field-lens and using the eye-lens alone as in the ocular, refocusing if necessary. Note also that the letters or other image is bordered by a colored haze (§ 22).

When looking into the ocular with the field-lens removed, the eye should not be held so close to the ocular, as the eye-point is considerably farther away than when the field-lens is in place (Pl. I, Fig. 7, and § 36).

§ 36. **The Eye-Point.**—This is the point above an ocular or simple microscope where the greatest number of emerging rays cross. Seen in profile, it may be likened to the narrowest part of an hour-glass. Seen in section (Pl. I, Fig. 7), it is the smallest and brightest light circle above the ocular. This is called the eye-point, for if the pupil of the eye is placed at this level, it will receive the greatest number of rays from the microscope, and consequently see the largest field.

Demonstrate the eye-point by having in position an objective and ocular as above (§ 35). Light the object brightly, focus the microscope, shade the ocular, then hold some ground-glass or a piece of the Japanese paper (§ 72) above the ocular and slowly raise and lower it until the smallest circle of light is found. By using different oculars it will be seen that the eye-point is nearer the eye-lens in low than in high oculars.

LIGHTING AND FOCUSING.

FOCUSING.

§ 37. Focusing is mutually arranging an object and the microscope so that a clear image may be seen.

With a simple microscope (§ 4) either the object or the microscope or both may be moved in order to see the image clearly, but with the compound microscope the object more conveniently remains stationary on the stage, and the tube or body of the microscope is raised or lowered (Pl. II, Fig. 10).

In general, the higher the power of the whole microscope whether simple or compound, the nearer together must the object and objective be brought. With the compound microscope, the higher the objective, and the longer the body of the microscope, the nearer together must the object and the objective be brought. If the oculars are not par-focal (§ 48), the higher the magnification of the ocular, the nearer must object and objective be brought.

§ 38. **Working Distance.**—By this is meant the space between the simple microscope and the object, or between the front-lens of the compound microscope and the object, when the microscope is in focus. This working distance is always considerably less than the equivalent focal length of the objective. For example, the front-lens of a ¼th in., or 6 mm. objective would not be ¼th inch, or 6 millimeters from the object when the microscope is in focus, but considerably less than that distance. If there were no other reason than the limited working distance of high objectives, it would be necessary to use very thin cover glasses over the object. (See § 16, 17). If too thick covers are used, it may be impossible to get an objective near enough an object to get it in focus. For objects that admit of examination with high powers it is always better to use thin covers.

LIGHTING.

§ 39. Unmodified sunlight should not be employed except in special cases. North light is best and most uniform. When the sky is covered with white clouds the light is most favorable. The light should come from the left; but if it is necessary to face the window a vertical, adjustable screen between the face and the window is desirable. If artificial illumination must be employed, use a lamp that gives a brilliant and steady light (§ 57).

It is of the greatest importance and advantage for one who is to use the microscope for serious work that he should comprehend and appreciate thoroughly the various methods of illumination, and the special appearances due to different kinds of illumination.

Depending on whether the light illuminating an object traverses the object or is reflected upon it, and also whether the object is symetrically lighted, or lighted more on one side than the other, light used in microscopy is designated as *reflected and transmitted, axial and oblique.*

§ 40. **Reflected, Incident or Direct Light.**—By this is meant light reflected upon the object in some way and then irregularly reflected from the object to the microscope. By this kind of light objects are ordinarily seen by the unaided eye, and the objects are mostly opaque. In Vertebrate Histology (Part II), reflected light is but little used; but in the study of opaque objects, like whole insects, etc., it is used a great deal. For low powers, ordinary daylight that naturally falls upon the object, or is reflected or condensed upon it with a mirror or con-

densing lens, answers very well. For high powers and for special purposes, special illuminating apparatus has been devised (Fig. 12). (See Carpenter).

§ 41. **Transmitted Light.**—By this is meant light which passes through an object from the opposite side. The details of a photographic negative are in many cases only seen or best seen by transmitted light, while the print made from it is best seen by reflected light (§ 40).

Almost all objects studied in Vertebrate Histology are lighted by transmitted light, and they are in some way rendered transparent or semi-transparent. The light traversing and serving to illuminate the object in working with a compound microscope is usually reflected from a plane or concave mirror, or from a mirror to an illuminator (§ 54), and thence transmitted to the object from below (Pl. II, Fig. 10, 13; Pl. III, Fig. 20).

§ 42. **Axial or Central Light.**—By this is understood light reaching the object, the rays of light being parallel to each other and to the optic axis of the microscope, or a diverging or converging cone of light whose axial ray is parallel with the optic axis of the microscope. In either case the object is symmetrically illuminated.

§ 43. **Oblique Light.** This is light in which parallel rays from a plane mirror form an angle with the optic axis of the microscope (Pl. II, Fig. 12, 13 c). Or if a concave mirror or a condenser is used, the light is oblique when the axial ray of the cone of light forms an angle with the optic axis (Pl. III, Fig. 20ª).

DIAPHRAGMS.

§ 44. **Diaphragms and their Proper Employment.**—Diaphragms are opaque disks with openings of various sizes, which are placed between the source of light or mirror and the object. In some cases an iris diaphragm is used, and then the same one is capable of giving a large range of openings. The object of a diaphragm, in general, is to cut off all adventitious light and thus to enable one to light the object in such a way that the light finally reaching the microscope shall all come from the object or its immediate vicinity.

§ 45. **Size and Position of Diaphragm Opening.**—The size of the opening in the diaphragm should be about that of the front lens of the objective used. For some objects and some objectives this rule may be quite widely departed from; one must learn by trial.

When lighting with a mirror the diaphragm should be as close as possible to the object in order, (a) that it may exclude all adventitious light from the object; (b) that it may not interfere with the most efficient illumination by the mirror by cutting off a part of the illuminating pencil. If the diaphragm is a considerable distance below the object, (1) it allows considerable adventitious light to reach the object and thus injures the distinctness of the microscopic image; (2) it prevents the use of very oblique light unless it swings with the mirror; (3) it cuts off a part of the illuminating cone from a concave mirror (Fig. 11).

With an illuminator (Pl. III, Fig. 20), the diaphragm serves to narrow the pencil to be transmitted through the condenser, and thus to limit the aperture or for any special purpose to be served (see § 61). Furthermore, by making the diaphragm opening excentric, oblique light may be used, or by using a diaphragm with a slit around the edge (central stop diaphragm), the center remaining opaque, the object may be lighted with a hollow cone of light all of the rays having great obliquity. In this way the so-called dark-ground illumination may be produced (§ 60; Pl. III, Fig. 20).

LIGHTING AND FOCUSING: EXPERIMENTS.

§ 46. **Lighting with a Mirror.**—Put a mounted fly's wing (see Ch. V, under mounting), under the microscope, put the ¾ in. (18 mm.) or other low objective in position, also a low ocular. With the coarse adjustment (Fig. 10), lower the body of the microscope within about 1 cm. of the object. Use an opening in the diaphragm about as large as the front lens of the objective; then with the plane mirror try to reflect light up through the diaphragm upon the object. One can tell when the field (§ 33) is illuminated, by looking at the object on the stage, but more satisfactorily by looking into the microscope. It sometimes requires considerable manipulation to light the field well. After using the plane side of the mirror turn the concave side into position and light the field with it. As the concave mirror condenses the light, the field will look brighter with it than with the plane mirror. Is it especially desirable to remember that the excellence of lighting depends in part on the position of the diaphragm (§ 45). If the greatest illumination is to be obtained from the concave mirror, its position must be such that its focus will be at the level of the object. This distance can be very easily determined by finding the focal point of the mirror in full sunlight.

§ 47. **Use of the Plane and of the Concave Mirror.**—The mirror should be freely movable, and have a plane and a concave face. The concave face is used when a large amount of light is needed, the plane face when a moderate amount is needed or when it is necessary to have parallel rays or to know the direction of the rays.

§ 48. **Focusing with Low Objectives.**—Place a mounted fly's wing under the microscope; put the three-fourths (18 mm.) objective in position, and also the lowest ocular. Select the proper opening in the diaphragm and light the object well with transmitted light (§41, 48).

Hold the head at about the level of the stage, look toward the window, and between the object and the front of the objective; with the coarse adjustment lower the body (Fig. 10), until the objective is within about half a cm. of the object. Then look into the microscope and slowly elevate the body with the coarse adjustment. The image will appear dimly at first, but will become very distinct by turning the body still higher. If the body is raised too high the image will become indistinct, and finally disappear. It will again appear if the body is lowered the proper distance.

When the microscope is well focused try both the concave and the plane mirrors, in various positions and note the effect. Put a high ocular in place of the low one (§ 27, 29, 31). If the oculars are not par-

focal it will be necessary to lower the tube somewhat to get the image in focus.*

Pull out the draw-tube (Fig. 10) 4–6 cm., thus lengthening the body of the microscope, and it will be found necessary to lower the tube of the microscope somewhat.

§ 49. **Pushing in the Draw-Tube.**—To push in the draw-tube, grasp the large milled ring of the ocular with one hand, and the milled head of the coarse adjustment with the other, and gradually push the draw-tube into the tube. If this were done without these precautions the objective might be forced against the object and the ocular thrown out by the compressed air.

§ 50. **Focusing with High Objectives.**—Employ the same object as before, elevate the body of the microscope and remove the ¾ in. (18 mm.) objective as indicated. Put the ⅕ in., (5 mm.) or a higher objective in place, and use a low ocular.

Light well, and employ the proper opening in the diaphragm, etc. (§ 45). Look between the front of the objective and the object as before (§ 48), and lower the body with the coarse adjustment till the objective almost touches the cover-glass over the object. Look into the microscope, and, with the coarse adjustment, raise the body very slowly until the image begins to appear, then turn the milled head of the fine adjustment (Fig. 10), first one way and then the other, if necessary, until the image is sharply defined.

Note that this high objective must be brought nearer the object than the low one, and that by changing to a higher ocular, if the oculars are not par-focal, or lengthening the body it will be found necessary to bring the objective still nearer the object, as with low objective (§ 48).

§ 51. **Always Focus Up**, as directed above. If one lowers the body only when looking at the end of the objective as directed above, there will be no danger of bringing the objective in contact with the object, as may be done if one looks into the microscope and focuses down.

When the instrument is well focused, move the object around in order to bring different parts into the field of view (§ 33). It may be necessary to re-focus with the fine adjustment every time a different part is brought into the field. In practical work, one hand is kept on the fine adjustment constantly, and the focus is continually varied.

---

* Par-focal oculars are so constructed, or so mounted, that those of different powers may be interchanged without the microscopic image becoming wholly out of focus. When high objectives are used, while the image may be seen after changing oculars, the instrument nearly always needs slight focusing. With low powers this may not be necessary.

CENTRAL AND OBLIQUE LIGHT WITH A MIRROR.

§ 52. **Axial Light,** (§ 42).—Place a preparation containing minute air-bubbles under the microscope. (The preparation may be easily made by beating a drop of mucilage on a slide and covering it. See Ch. II). Use a ⅛ inch (3 mm.) or No. 7 objective, and a medium ocular. Remove all the diaphragms and the sub-stage. Focus the microscope and select a very small bubble, one whose image appears about 1 mm. in diameter, then arrange the plane mirror so that the light spot in the bubble appears exactly in the center. Without changing the position of the mirror in the least, replace the air-bubble preparation by one of *Pleurasigma angulatum* or some other finely marked diatom. Study the appearance very carefully.

§ 53. **Oblique Light,** (§ 43).—Swing the mirror far to one side so that the rays reaching the object may be very oblique to the optic axis of the microscope. Study carefully the appearance of the diatom with the oblique light. Compare the different appearance with that of central light. The effect of oblique light is not so striking with histological preparations as with diatoms.

It should be especially noted in §§ 52, 53, that one cannot determine the exact direction of the rays by the position of the mirror. This is especially true for axial light, (§ 52). To be certain that the light is axial some such test as that given in § 52 should be applied. (See also Ch. II, under Air-bubbles).

ABBE ILLUMINATOR OR CONDENSER.

§ 54. For all powers, but especially for high power objectives, a condenser or illuminator is of great advantage. The one most generally useful was designed by Abbe. It consists of two or three very large lenses which are placed in some form of mounting beneath the stage. It serves to concentrate a very wide pencil of light from the mirror upon the object. For the best work in modern histology the Abbe illuminator is almost as indispensable as the homogeneous immersion objectives (Pl. III, Fig. 20).

§ 55. **Centering and Arrangement of the Illuminator.**—The proper position of the illuminator for high objectives is one in which the beam of light traversing it is brought to a focus on the object. If parallel rays are reflected from the plane mirror to it, they will be focused only a few millimeters above the upper lens of the illuminator; consequently the illuminator should be about on the level of the top of the stage and therefore almost in contact with the lower surface of the slide. For some purposes, when it is desirable to avoid the loss of light by reflection or refraction, a drop of water or homogeneous immersion

fluid is put between the slide and condenser, forming the so-called immersion illuminator. This is necessary only with objectives of high power and large aperture or for dark-ground illumination.

Centering the Illuminator.—The illuminator should be centered to the optic axis of the microscope, that is the optic axis of the condenser and of the microscope should coincide. If one has a pin-hole diaphragm to put over the end of the condenser (Fig. 20)—that is a diaphragm with a small central hole—the central opening should appear to be in the middle of the field of the microscope. If it does not, the condenser should be moved from side to side by loosening the centering screws until it is in the center of the field. In case no pin-hole diaphragm accompanies the condenser, one may put a very small drop of ink, as from a pen-point, on the center of the upper lens and look at it with the microscope to see if it is in the center of the field. If it is not, the condenser should be adjusted until it is. The microscope and illuminator axes may not be entirely coincident even when the center of the upper lens appears in the center of the field, as there may be some lateral tilting of the condenser, but the above is the best the ordinary worker can do, and unless the mechanical arrangements of the illuminator are very deficient, it will be very nearly if not absolutely centered.

§ 56. **Mirror and Light for the Abbe Illuminator.**—It is best to use daylight for this as for all other means of illumination. The rays of daylight are practically parallel, and it is best, therefore, to employ the plane mirror for all but the lowest powers. If low powers are used the whole field might not be illuminated with the plane mirror and the condenser close to the object; furthermore, the image of the window frame, objects outside the building, as trees, etc., would appear with unpleasant distinctness in the field of the microscope. To overcome these defects, one can lower the condenser and thus light the object with a diverging cone of light, or use the concave mirror and attain the same end when the condenser is close to the object (Pl. III, Fig. 20).

§ 57. **Lamplight.**—If one must use lamplight, it is recommended that a large condensing lens be placed in such a position between the light and the mirror that a picture of the lamp flame is thrown upon the mirror. If one does not have a condensing lens the concave mirror may be used to render the rays less divergent. It may be necessary to lower the illuminator somewhat in order to illuminate the object in its focus.

ABBE ILLUMINATOR : EXPERIMENTS.

§ 58. **Abbe Illuminator, Axial and Oblique Light.**—Use a diaphragm a little larger than the front lens of the $\frac{1}{8}$ (3 mm.) or No. 7

objective, have the illuminator on the level or nearly on the level of the upper surface of the stage, and use the plane mirror. Be sure that the diaphragm carrier is in the notch indicating that it is central in position. Use the *Pleurasigma* as object. Study carefully the appearance of the diatom with this central light, then make the diaphragm excentric so as to light with oblique light. The differences in appearance will probably be even more striking than with the mirror alone (§§ 52, 53).

§ 59. **Lateral Swaying of the Image.**—Frequently in studying an object, especially with a high power, it will appear to sway from side to side in focusing up or down. A glass stage micrometer or fly's wing is an excellent object. Make the light central or axial and focus up and down and notice that the lines simply disappear or grow dim. Now make the light oblique, either by making the diaphragm opening excentric or if simply a mirror is used, by swinging the mirror sidewise. On focusing up and down, the lines will sway from side to side. What is the direction of apparent movement in focusing down with reference to the illuminating ray? What in focusing up? If one understands this experiment it may sometime save a great deal of confusion.

§ 60. **Dark-Ground Illumination.**—When an object is lighted with rays of a greater obliquity than can get into the front-lens of the objective, the field will appear dark (Pl. III, Fig. 22). If now the object is composed of fine particles, or is semi-transparent, it will refract or reflect the light which meets it, in such a way that a part of the very oblique rays will pass into the objective, hence as light reaches the objective only from the object, all the surrounding field will be dark and the object will appear like a self-luminous one on a dark back ground. This form of illumination is only successful with low powers and objectives of small aperture. It is well to make the illuminator immersion for this experiment, see § 55.

(A) *With the Mirror.*—Remove all the diaphragms so that very oblique light may be used, employ a stage micrometer in which the lines have been filled with graphite, use a ¾ths (18 mm.) or No. 4 objective, and when the light is sufficiently oblique the lines will appear something like streaks of silver on a black back-ground. A specimen like that described below in (B) may also be used.

(B) *With the Abbe Illuminator.*—Have the illuminator so that the light would be focused on the object (see above § 55) and use a diaphragm with the slit opening; employ the same objective as in (A). For object place a drop of a 10% solution of salicylic acid in alcohol on the middle of a slide and allow it to dry and crystallize. The crystals will appear brilliantly lighted on a dark back-ground. Put in an ordinary diaphragm and make the light oblique by making the diaphragm eccentric. The same specimen may also be tried with a

mirror and oblique light. In order to appreciate the difference between this dark ground and ordinary transmitted light illumination, use an ordinary diaphragm and observe the crystals.

A very striking and instructive experiment may be made by adding a very small drop of the solution to the dried preparation, putting it under the microscope very quickly, lighting for dark-ground illumination and then watching the crystallization.

REFRACTION AND COLOR IMAGES.

§ 61. **Refraction Images** are those mostly seen in studying microscopic objects. They are the appearances due to the refraction of the light in passing from the mounting medium into the object and from the object back into the mounting medium. With such images the diaphragms should not be too large (see § 45).

If the color and refractive index of the object were exactly like the mounting medium it could not be seen. In most cases both refractive index and color differ somewhat, there is then a combination of color and refraction images which is a great advantage. This method of illumination is mostly used in histology.

§ 62. **Color Images.**—These are images of objects which are strongly colored and lighted with so wide an aperture that the refraction images are drowned in the light. Such images are obtained by removing the diaphragm or by using a larger opening. This method of illumination is specially applicable to the study of stained microbes. (See below § 67).

ADJUSTABLE, WATER AND HOMOGENEOUS OBJECTIVES.

EXPERIMENTS.

§ 63. **Adjustment for Objectives.**—As stated above (§ 16), the aberration produced by the cover-glass (Pl. V., Fig. 41), is compensated for by giving the combinations in the objective a different relative position than they would have if the objective were to be used on uncovered objects. Although this relative position cannot be changed in unadjustable objectives, one can secure the best results of which the objective is capable by selecting covers of the thickness for which the objective was corrected. (See table in § 17). Adjustment may be made also by *increasing* the tube-length (§ 17) for covers *thinner* than the standard, and by *shortening* the tube-length for covers *thicker* than the standard (§ 17).

Adjustable Objectives.—The proper adjustment of objectives, that is, the adjustment which gives the truest image, requires both insight and experience; for the structure of an object does not appear the same with different adjustments of the objective. And as the opinion of different observers on the structure of objects varies, they adjust the objectives differently, and try to obtain the adjustment which will show a structure in accordance with their opinion. Eyes also differ, and two observers might find it necessary to adjust the same objective differently to produce an identical appearance for each of them.

In learning to adjust objectives, it is best for the student to choose some object whose structure is well agreed upon, and then to practice lighting it, shading the stage and adjusting the objective, until the proper appearance is obtained. The adjustment is made by turning a ring or collar which acts on a screw and increases or diminishes the distance between the systems of lenses, usually the front and the back systems (Fig. 13). In adjustable objectives the back systems should be movable, the front one remaining fixed so that there will be no danger of bringing the objective down upon the object. If the front system is movable, the body of the microscope should be raised slightly every time the adjustment is altered.

General Directions.—(A) The thinner the cover-glass the further must the systems of lenses be separated, *i. e.*, the adjusting collar is turned nearer the zero or the mark "uncovered," and conversely, (B) the thicker the cover-glass, the closer together are the systems brought by turning the adjusting collar *from* the zero mark. This also increases the magnification of the objective (Ch. III).

The following specific directions for making the cover-glass adjustment are given by Mr. Wenham [(C. 166)] : "Select any dark speck or opaque portion of the object, and bring the outline into perfect focus ; then lay the finger on the milled-head of the fine motion, and move it briskly backwards and forwards in both directions from the first position. Observe the expansion of the dark outline of the object, both when within and when without the focus. If the greater expansion or coma, is when the object is *without* the focus, or farthest from the objective [*i. e.*, in focusing up], the lenses must be placed further asunder, or toward the mark uncovered [*i. e.*, the adjusting collar is turned toward the zero mark as the cover-glass is too thin for the present adjustment]. If the greater expansion is when the object is within the focus, or nearest the objective, [*i. e.*, in focusing down], the lenses must be brought closer together or toward the mark covered [*i. e.*, the adjusting collar should be turned away from the zero mark, the cover-glass being too thick for the present adjustment]." *In most objectives the collar is graduated arbitrarily, the zero (O) mark representing the position for uncovered objects. Other objectives have the collar graduated to correspond to the various thickness of cover-glasses for which the objective may be adjusted. This seems to be an admirable plan ; then if one knows the thickness of the cover-glass on the preparation (Ch. V) the adjusting collar may be set at a corresponding mark, and one will feel confident that the adjustment will be approximately eorrect. It is then only necessary for the observer to make the slight adjustment to compensate for the mounting medium or any variation from the standard length of the tube of the microscope.* In adjusting for variations of the length of the tube from

the standard it should be remembered that: (**A**) If the tube of the microscope is longer than the standard for which the objective was corrected, the effect is approximately the same as thickening the cover-glass, and therefore the systems of the objective must be brought closer together, *i. e.*, the adjusting collar must be turned *away from* the zero mark. (**B**) If the tube is shorter than the standard for which the objective is corrected, the effect is approximately the same as diminishing the thickness of the cover-glass, and the systems must therefore be separated, that is, the adjusting collar turned *toward* the zero mark (Fig. 57).

Furthermore, whatever the interpretation by different opticians of what should be included in "tube-length," and the exact length in millimeters, its importance is very great; for each objective gives the most perfect image of which it is capable with the "tube-length" for which it is corrected, and the more perfect the objective the greater the ill effects on the image of varying the "tube-length" from this standard. The plan of designating exactly what is meant by "tube-length," and engraving on each objective the "tube-length" for which it is corrected, is to be commended, for it is manifestly difficult for each worker with the microscope to find out for himself for what "tube-length" each of his objectives was corrected.

§ 64. **Water Immersion Objectives.**—Put a water immersion objective in position (§ 30) and the fly's wing for object under the microscope. Place a drop of distilled water on the cover-glass and with the coarse adjustment lower the tube till the objective dips into the water, then light the field well and turn the fine adjustment one way and the other till the image is clear. Water immersions are exceedingly convenient in studying the circulation of the blood and for many other purposes where aqueous liquids are liable to get on the cover-glass. If the objective is adjustable, follow the directions given in § 63.

When one is through using a water immersion objective, remove it from the microscope and with some Japanese bibulous paper (§ 72) wipe all the water from the front-lens. Unless this is done dust collects and sooner or later the front-lens would be clouded. It is better to use distilled water to avoid the gritty substances that are liable to be present in natural waters, as these gritty particles might scratch the front-lens.

HOMOGENEOUS IMMERSION OBJECTIVES: EXPERIMENTS.

§ 65. As stated above, these are objectives in which a liquid of the same refractive index as the front-lens of the objective is placed between the front-lens and the cover-glass.

Put a 2 mm. ($\frac{1}{12}$th in.) homogeneous immersion objective in position, employ an Abbe illuminator.

§ 66. **Refraction Images.**—Use some histological specimen like a muscular fiber as object, make the diaphragm opening only slightly larger than the front-lens, add a drop of the homogeneous immersion liquid and focus as directed in §§ 64 and 67. The object will be clearly seen in all its details by the unequal refraction of the light traversing it. The difference in color between it and the surrounding medium will also increase the sharpness of the outline. If an air bubble preparation (§ 52) were used, one would get pure, refraction images.

§ 67. **For Color Images.**—Use some stained microbes, as *Bacillus tuberculosis* for object. Put a drop of the immersion liquid on the cover-glass or the front-lens of the homogeneous objective. Remove the diaphragms from the illuminator or use a very large opening. Focus the objective down so that the immersion fluid is in contact with both the front-lens and the cover-glass, then with the fine adjustment get the microbes in focus. They will stand out as colored rods on a bright field.

§ 68. **Shading the Object.**—To get the clearest image of an object no light should reach the eye except from the object. A handkerchief or a dark cloth wound around the objective will serve the purpose. Often the proper effect may be obtained by simply shading the top of the stage with the hand or with a piece of bristol board. Unless one has a very favorable light the shading of the object is of the greatest advantage, especially with homogeneous immersion objectives.

§ 69. **Cleaning Homogeneous Objectives.**—After one is through with a homogeneous objective, it should be carefully cleaned as follows: Wipe off the homogeneous liquid with a piece of the Japanese paper (§ 72), then if the fluid is cedar oil, wet one corner of a fresh piece in benzin and wipe the front-lens with it. Immediately afterward wipe with a dry part of the paper. The cover-glass of the preparation can be cleaned in the same way. If the homogeneous liquid is a glycerin mixture proceed as above, but use water instead of benzin to remove the last traces of glycerin.

CARE OF THE MICROSCOPE.

§ 70. The microscope should be handled carefully, and kept perfectly clean. The oculars and objectives should never be allowed to fall.

When not in use keep it in a place as free as possible from dust.

All parts of the microscope should be kept free from liquids, especially from acids, alkalies, alcohol, benzin, turpentine and chloroform.

§ 71. **Care of the Mechanical Parts.**—To clean the mechanical

parts put a small quantity of some fine oil, (olive oil and benzin equal parts), on a piece of chamois leather or on the Japanese paper, and rub the parts well, then with a clean dry piece of the chamois or paper wipe off the oil. If the mechanical parts are kept clean in this way a lubricator is rarely needed. Where opposed brass surfaces cut, a very slight application of equal parts of beeswax and tallow well melted together serves a good purpose.

In cleaning lacquered parts, benzin alone answers well, but it should be quickly wiped off with a clean piece of the Japanese paper. Do not use alcohol as it dissolves the lacquer.

§ 72. **Care of the Optical Parts.**—These must be kept scrupulously clean in order that the best results may be obtained.

Glass surfaces should never be touched with the fingers, for that will soil them.

The glass of which the lenses are made is quite soft, consequently it is necessary that only soft, clean cloths or paper be used in wiping them.

"*Paper for Cleaning the Lenses of Objectives and Oculars.*—For the last six years the so-called Japanese filter paper (the bibulous paper often used by dentists when filling teeth), has been used in the laboratory for cleaning the lenses of oculars and objectives, and especially for removing the fluid used with immersion objectives. Whenever a piece is used once it is thrown away. It has proved more satisfactory than cloth or chamois, because dust and sand are not present; and from its bibulous character it is very efficient in removing liquid or semi-liquid substances."

Dust may be removed with a camel's hair brush, or by wiping with the soft paper.

Cloudiness may be removed from the glass surfaces by breathing on them, then wiping quickly with a soft cloth or the bibulous paper.

Cloudiness on the inner surfaces of the ocular lenses may be removed by unscrewing them and wiping as directed above. A high objective should never be taken apart by an inexperienced person.

If the cloudiness cannot be removed as directed above, moisten one corner of the cloth or paper with 95 per cent. alcohol, wipe the glass first with this, then with the dry cloth or the paper.

Water may be removed with soft cloth or the paper.

Glycerin may be removed with cloth or paper saturated with distilled water; remove the water as above.

Blood or other albuminous material may be removed while fresh with a moist cloth or paper, the same as glycerin. If the material has dried to the glass, it may be removed more readily by adding a small quantity of ammonia to the water in which the cloth is moistened, (water 100 cc., ammonia 1 cc).

Canada Balsam, damar, paraffin, or any oily substance, may be removed with a cloth or paper wet with chloroform, turpentine or benzin. The application of these liquids and their removal with a soft, dry cloth or paper should be as rapid as possible, so that none of the liquid will have time to soften the setting of the lenses.

Shellac Cement may be removed by the paper or a cloth moistened in 95 per cent. alcohol.

Brunswick Black, *Gold Size*, and all other substances soluble in chloroform, etc., may be removed as directed for balsam and damar.

In general, use a solvent of the substance on the glass and wipe it off quickly with a fresh piece of the Japanese paper.

It frequently happens that the upper surface of the back combination of the objective becomes dusty. This may be removed in part by a brush, but more satisfactorily by using a piece of the soft paper loosely twisted. When most of the dust is removed some of the paper may be put over the end of a pine stick (like a match stick) and the glass surface carefully wiped.

CARE OF THE EYES.

§ 73. Keep both eyes open, using the eye-screen if necessary (Fig. 18, and Pl. II, Fig. 16); and divide the labor between the two eyes, *i. e.*, use one eye for observing the image awhile and then the other. In the beginning it is not advisable to look into the microscope continuously for more than half an hour at a time. One never should work with the microscope after the eyes feel fatigued. After one becomes accustomed to microscopic observation he can work for several hours with the microscope without fatiguing the eyes.

FIG. 18.—*Ward's Eye-Shade.*

This is due to the fact that the eyes become inured to labor like the other organs of the body by judicious exercise. It is also due to the fact that but very slight accommodation is required of the eyes, the eyes remaining nearly in a condition of rest as for distant objects. The fatigue incident upon using the microscope at first is due partly at least to the constant effort to accommodate the eye for a near object. With a microscope of the best quality, and suitable light—that is light which is steady and not so bright as to dazzle the eyes nor so dim as to strain them in determining details— microscopic work should improve rather than injure the sight.

A LABORATORY COMPOUND MICROSCOPE.

§ 74. For the purpose of modern histological investigation and of advanced microscopical work in general, a microscope should have something like the following character: Its optical outfit should comprise, (a) dry objectives of 50 mm. (2

in.), 16–18 mm. (¾ in.) and 3 mm. (⅛ in.) equivalent focus. There should be present also a 2 mm. ($\frac{1}{12}$ in.) or 1.5 mm. ($\frac{1}{16}$ in.) homogeneous immersion objective. Of oculars there should be several of different power. An Abbe illuminator, and an Abbe camera lucida are also necessities. A micro-spectroscope and a micro-polarizer are very desirable.

Even in case all the optical parts cannot be obtained in the beginning, it is wise to secure a stand upon which all may be used when they are finally secured.

Mechanical Parts or Stand.—The stand should be low enough so that it can be used in a vertical position on an ordinary table without inconvenience; it should have a jointed (flexible) pillar for inclination at any angle to the horizontal. The adjustments for focusing should be two,—a coarse adjustment or rapid movement with rack and pinion, and a fine adjustment by means of a micrometer screw. Both adjustments should move the entire body of the microscope. The body or tube should be short enough for objectives corrected for the short or 160 millimeter tube-length, and the draw-tube should be graduated in centimeters and millimeters. The lower end of the draw-tube and of the tube should each possess a standard screw for objectives (Pl. II, Fig. 10). The stage should be quite large for the examination of slides with serial sections; it is also of considerable advantage to have the stage with a circular, revolving top, and two centering screws with milled heads. In this way a mechanical stage with limited motion is secured, and this is of the highest advantage in using powerful objectives. The sub-stage fittings should be so arranged as to enable one to dispense entirely with diaphragms, to use ordinary diaphragms, or to use the Abbe illuminator. The illuminator mounting should allow up and down motion, preferably by rack and pinion. The base should be sufficiently heavy and so arranged that the microscope will be steady in all positions, and interfere the least possible amount with the manipulation of the mirror and other sub-stage accessories.

A microscope with all the accessories mentioned above would cost about $170 in Europe. Without the micro-spectroscope and micro-polarizer about $115. To get the American price of a foreign instrument or the catalogue price of one of American manufacture, the amount of duty (*i. e.*, 60 per cent.) should be added.

REFERENCES FOR CHAPTER I.

In the appendix will be given a bibliography, with full titles, of the works and periodicals referred to. In the text the works are mostly designated by single letters, and the periodicals by the initial letters of their titles.

The works of Beale (B.), Bausch (Bau.), Behrens (Beh.), Carpenter (C.), Dippel (D.), Frey (F.), Hogg (H.), Mayall, Nägeli und Schwendener, Robin.

The following special articles in periodicals may be examined with advantage:

§ 14. Apochromatic Objectives, etc. Dippel in Zeit. wiss. Mikr. 1886, p. 303; also in the Jour. Roy. Micr. Soc., 1886, pp. 316, 849, 1110; same, 1890, p. 480; Zeit. f. Instrumentenk., 1890, pp. 1-6; Micr. Bullt., 1891, pp. 6–7.

§ 17. Tube-length, etc. Gage, Proc. Amer. Soc. Micrs., 1887, pp. 168-172; also in the Microscope, the Jour. Roy. Micr. Soc., and in Zeit. wiss. Mikr., 1887-8, Bausch, Proc. Amer. Soc. Micrs., 1890, pp. 43-49; also in the Microscope, 1890, pp. 289-296.

§ 18. Aperture. J. D. Cox, Presidential Address, Proc. Amer. Soc. Micrs., 1884, pp. 5-39, Jour. Roy. Micr. Soc., 1881, pp. 303, 348, 365, 388; 1882, pp. 300, 460; 1883, p. 790; 1884, p. 20.

§ 54. The Abbe Illuminator. Archiv f. mikr. Anat., Vol. IX, (1873) p. 469.

CHAPTER II.

INTERPRETATION OF APPEARANCES.

APPARATUS AND MATERIAL FOR CHAPTER II.

A laboratory, compound microscope (§§ 74); Preparation of fly's wing (see appendix); 50 per cent. glycerin; Slides and covers; Preparation of letters in stairs (Fig. 23 in Pl. III); Mucilage for air-bubbles and olive or clove oil for oil-globules (§ 81); Solid glass rod, and glass tube (§ 89); Collodion (§ 91); Carmine, India ink, or lamp black (§ 93); Frog, and castor oil or paraffin, micro-polariscope (§ 95).

INTERPRETATION OF APPEARANCES UNDER THE MICROSCOPE.

§ 75. **General Remarks.**—The experiments in this chapter are given secondarily for drill in manipulation, but primarily so that the student may not be led into error or puzzled by appearances which are constantly met with in microscopical investigation. Any one can look into a microscope, but it is quite another matter to interpret correctly the meaning of the appearances seen.

It is especially important to remember that the more of the relations of any object are known, the truer is the comprehension of the object. In microscopical investigation every object should be scrutinized from all sides and under all conditions in which it is likely to occur in nature and in microscopical investigation. It is best also to begin with objects of considerable size whose character is well known, to look at them carefully with the unaided eye so as to see them as wholes and in their natural setting. Then a low power is used, and so on step by step until the highest power available has been employed. One will in this way see less and less of the object as a whole, but every increase in magnification will give increased prominence to detail, detail which might be meaningless when taken alone and independent of the object as a whole. The pertinence of this advice will be appreciated when the student undertakes to solve the problems of histology; for even after all the years of incessant labor spent in trying to make out the structure of man and the lower animals, many details are still in doubt, the same visual appearances being quite differently interpreted by eminent observers.

§ 76. **Dust or Cloudiness on the Ocular.**—Employ the 18 mm. (¾ in.) objective, low ocular, and fly's wing, as object.

Unscrew the field-lens and put some particles of lint or dark cloth on its upper surface. Replace the field-lens and put the ocular in position

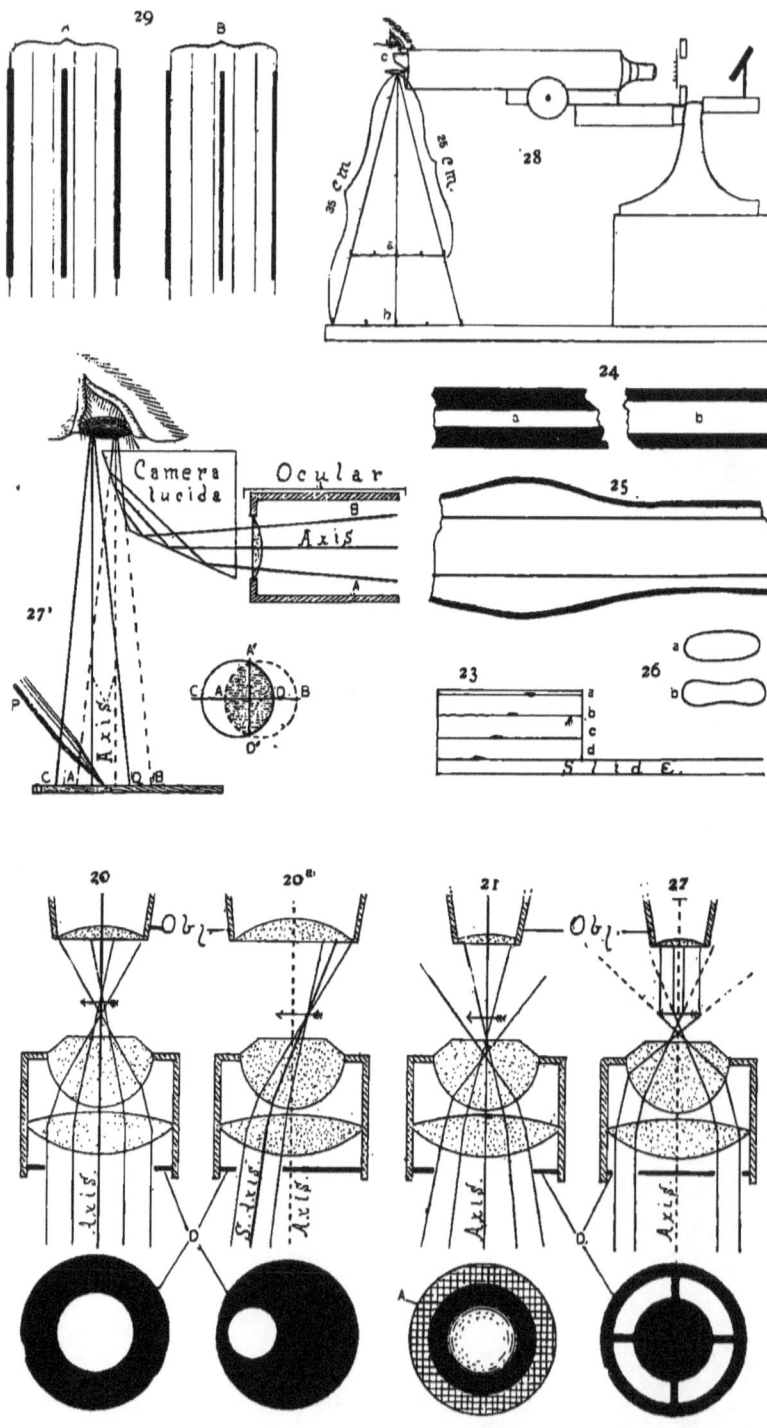

EXPLANATION OF PLATE III.

Fig. 20, 20ᵃ, 21 and 22. Sectional views of the Abbe Illuminator of 1.20 N. A. (§§ 18, 54), showing various methods of illumination (§§ 54–59). Fig. 20, axial light with parallel rays. Fig. 20ᵃ, oblique light. Fig. 21, axial light with converging beam. Fig. 22, dark-ground illumination with a central stop diaphragm.

Axis. The optic axis of the illuminator and of the microscope. The illuminator is centered, that is its optic axis is a prolongation of the optic axis of the microscope.

S. Axis. Secondary axis. In oblique light the central ray passes along a secondary axis of the illuminator, and is therefore oblique to the principal axis.

A. Fig. 21 represents the upper part of the illuminator.

D D. *Diaphragms.* These are placed in sectional and in face views. The diaphragm is placed between the mirror and the illuminator. In Fig. 20 the opening is excentric for oblique light, and in Fig. 22 the opening is a narrow band, the central part being stopped out, and thus giving rise to dark-ground illumination (§ 60).

Obj. Obj. The front of the objective.

Fig. 23. Showing the method of mounting letters in stairs to show the order of coming into focus.

a, b, c, d. The various letters indicated by the oblique row of black marks in the sectional view.

Slide. The glass slide on which the letters are mounted.

Fig. 24. Glass rod showing the appearance in air (a), and in 50 per cent. glycerin (b), (§ 80).

Fig. 25. Glass rod coated with collodion to show Double Contour. At the left the collodion is represented as collecting in a drop (§ 91).

Fig. 26. Mammalian blood-corpuscles on edge to show a surface view (a), and an optical section (b).

Fig. 27. *Wollaston's Camera Lucida*, showing the rays from the microscope and from the drawing surface, and the position of the pupil of the eye.

Axis, Axis. Axial rays from the microscope and from the drawing surface. (§ 121)

Camera Lucida. A section of the quadrangular prism showing the course of the rays in the prism from the microscope to the eye. As the rays are twice reflected, they have the same relation on entering the eye that they would have by looking directly into the ocular.

A B. The lateral rays from the microscope and their projection on the drawing surface.

C D. Rays from the drawing surface to the eye.

A D, A' D'. Overlapping portion of the two fields, where both the microscopic image and the drawing surface, pencil, etc., may both be seen. It is represented by the shaded part in the overlapping circles at the right.

Ocular. The ocular of the microscope.

P. The drawing pencil. Its point is shown in the overlapping fields.

Fig. 28. Figure showing the position of the microscope, the camera lucida, and the eye, and the different sizes of the image depending upon the distance at which it is projected from the eye. (a) The size at 25 cm.; (b) at 35 cm. (§ 104).

Fig. 29. Figure showing the appearance of the lines of the stage micrometer (the coarse lines) and of the ocular micrometer when using a high objective (§ 117).

A. One method of measuring the spaces by putting the ocular micrometer line opposite the center of the stage micrometer line.

B. Method of measuring the space of the stage micrometer by putting one line of the ocular micrometer at the *inside* and one at the *outside* of the lines of the stage micrometer (§ 117).

(§ 31). Light the field well and focus sharply. The image will be clear, but part of the field will be obscured by the irregular outline of the particles of lint. Move the object to make sure this appearance is not due to it.

Grasp the ocular by the milled ring, just above the tube of the microscope, and rotate it. The irregular object will rotate with the ocular. Cloudiness or particles of dust on any part of the ocular, may be detected in this way.

§ 77. **Dust or Cloudiness on the Objective.**—Employ the same ocular and objective as before (§ 76), and the fly's wing as object. Focus and light well, and observe carefully the appearance. Rub glycerin on one side of a slide near the end. Hold the clean side of this end close against the objective. The image will be obscured, and can not be made clear by focusing. Then use a clean slide, and the image may be made clear by elevating the body slightly. The obscurity produced in this way is like that caused by clouding the front-lens of the objective. Dust would make a dark patch on the image that would remain stationary while the object or ocular was moved.

If too small a diaphragm is employed, only the central part of the field will be illuminated, and around the small light circle will be seen a dark ring.

§ 78. **Relative Position of Objects or parts of the same object.**— The general rule is that objects highest up come into focus *last* in focusing up, *first* in focusing down.

§ 79. **Objects Having Plane or Irregular Outlines.**—As object use three printed letters mounted in stairs in Canada balsam (Pl. III, Fig. 23, Ch. V). The first letter is placed directly upon the slide, and covered with a small piece of glass about as thick as a slide. The second letter is placed upon this and covered in like manner. The third letter is placed upon the second thick cover and covered with an ordinary cover-glass. The letters should be as near together as possible, but not overlapping. Employ the same ocular and objective as above (§ 76).

Lower the tube till the objective almost touches the top letter, then look into the microscope, and slowly focus up. The lowest letter will first appear, and then, as it disappears, the middle one will appear, and so on. Focus down, and the top letter will first appear, then the middle one, etc. The relative position of objects is determined exactly in this way in practical work.

§ 80. **Transparent Objects Having Curved Outlines.**—The success of these experiments will depend entirely upon the care and skill used in preparing the objects, in lighting, and in focusing.

Employ a 5 mm. (⅕ in.) or higher objective and a high ocular for all

the experiments. It may be necessary to shade the object (§ 66) to get satisfactory results. When a diaphragm is used the opening should be small (§ 44).

§ 81. **Air Bubbles.**—Prepare these by placing a drop of thin mucilage on the center of a slide and beating it with a scalpel blade until the mucilage looks milky from the inclusion of air bubbles. Put on a coverglass (Ch. V), but do not press it down.

§ 82. **Air Bubbles with Central Illumination.**—Shade the object; and with the plane mirror, light the field with central light (Pl. II, Fig. 13, § 42).

Search the preparation until an air bubble is found appearing about 1 mm. in diameter, get it into the center of the field and if the light is central the air bubble will appear with a wide, dark, circular margin and a small bright center. If the bright spot is not in the center, adjust the mirror until it is.

This is one of the simplest and surest methods of telling when the light is central or axial (§ 52).

Focus both up and down, noting that in focusing up the central spot becomes very clear and the black ring very sharp. On elevating the body still more the center becomes dim, and the whole bubble loses its sharpness of outline.

§ 83. **Air Bubbles with Oblique Illumination.**—Remove the substage of the microscope (Fig. 10), and all the diaphragms. Swing the mirror so that the rays may be sent very obliquely upon the object (Fig. 13, C). The bright spot will appear no longer in the center but on the side *away from* the mirror (Fig. 19).

§ 84. **Oil Globules.**—Prepare these by beating a small drop of clove oil with mucilage on a slide and covering as directed for air bubbles (§ 81).

§ 85. **Oil Globules with Central Illumination.**—Use the same diaphragm and light as above (§ 82). Find an oil globule appearing about 1 mm. in diameter. If the light is central the bright spot will appear in the center as with air (§ 82). Focus up and down as with air; and note that the bright center of the oil globule is clearest *last* in focusing up.

§ 86. **Oil Globules with Oblique Illumination.**—Remove the substage, etc., as above, and swing the mirror to one side and light with oblique light. The bright spot will be eccentric, and will appear to be on the *same* side as the mirror (Fig. 19).

§ 87. **Oil and Air Together.**—Make a preparation exactly as described for air bubbles (§ 81), and add at one edge a little of the mixture of oil and mucilage (§ 84).; cover and examine.

The sub-stage need not be used in this experiment. Search the preparation until an air bubble and an oil globule, each about 1 mm. in diameter, are found in the same field of view. Light first with central light, and note that in focusing up the air bubble comes into focus first and that the central spot is smaller than that of the oil globule. Then, of course, the black ring will be wider in the air bubble than in the oil globule. Make the light oblique. The bright spot in the air bubble will move *away from* the mirror while that in the oil globule will move *toward* it. See Fig. 19.*

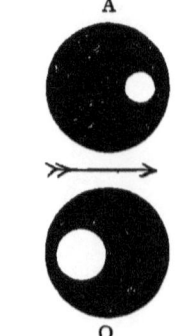

FIG. 19.—*Very Small Globule of Oil* (O) *and an Air-Bubble* (A) *Seen by Oblique Light. The Arrow Indicates the Direction of the Light Rays.*

§ 88. **Air and Oil by Reflected Light.**—Cover the diaphragm or mirror so that no transmitted light (§ 41) can reach the preparation, using the same preparation as in (§ 87). The oil and air will appear like globes of silver on a dark ground. The part that was darkest in each will be lightest, and the bright central spot will be somewhat dark.†

§ 89. **Distinctness of Outline.**—In refraction images (§§ 61, 66) this depends on the difference between the refractive power of a body and that of the medium which surrounds it. The oil and air were very distinct in outline as each differed greatly in refractive power from the medium which surrounded them, the oil being more refractive than the mucilage and the air less.

Place a fragment of a cover-glass on a clean slide, and cover it (see Ch. V, under mounting). The outline will be very distinct with the unaided eye. Use it as object and employ the 18 mm. (¾ in.) objective and high ocular. Light with central light. The fragment will be outlined by a dark band. Put a drop of water at the edge of the cover-glass. It will run in and immerse the fragment. The outline will remain distinct, but the dark band will be somewhat narrower. Remove the cover-glass, wipe it dry, and wipe the fragment and slide dry also. Put a drop of 50% glycerin on the middle of the slide and mount

---

* It should be remembered that the image in the compound microscope is inverted (Fig. 5), hence the bright spot really moves toward the mirror for air, and away from it for oil.

† It is possible to distinguish oil and air optically, as described above, only when quite high powers are used and very small bubbles are selected for observation. If an 18 mm. (¾ in.) is used instead of a 3 mm. (⅛ in.) objective, the appearances will vary considerably from that given above for the higher power. It is well to use a low as well as a high power. Marked differences will also be seen in the appearances with objectives of small and of large aperture.

the fragment of cover-glass in that. The dark contour will be much narrower than before.

Draw a solid glass rod out to a fine thread. Mount one piece in air, and the other in 50% glycerin. Put a cover-glass on each. Employ the same optical arrangement as before. Examine the one in air first. There will be seen a narrow, bright band, with a wide dark band on each side.

The one in glycerin will show a much wider bright central band, with the dark borders correspondingly narrow (Pl. III, Fig. 24).

If the glass rod or any other object were mounted in a medium of the same color and refractive power, it could not be distinguished from the medium.*

§ 90. **Highly Refractive.**—This expression is often used in describing microscopic objects, (medulated nerve fibres for example), and means that the object will appear to be bordered by a wide, dark margin when it is viewed by transmitted light. And from the above (§ 89), it would be known that the refractive power of the object, and the medium in which it was mounted must differ considerably.

§ 91. **Doubly Contoured.**—This means that the object is bounded by two, usually parallel dark lines with a lighter band between them. In other words the object is bordered by (1) a dark line, (2) a light band, and (3) a second dark line (Pl. III, Fig. 25).

This may be demonstrated by coating a fine glass rod (§ 89) with one or more coats of collodion or celloidin and allowing it to dry, and then mounting in 50% glycerin as above (§ 89). Employ a 5 mm. (⅕ in.) or higher objective, light with transmitted light, and it will be seen that where the glycerin touches the collodion coating there is a dark line—next this is a light band, and finally there is a second dark line where the collodion is in contact with the glass rod† (Pl. III, Fig. 25).

§ 92. **Optical Section.**—The appearance obtained in examining transparent or nearly transparent objects with a microscope when some plane below the upper surface of the object is in focus. The upper part of the object which is out of focus obscures the image but slightly. By changing the position of the objective or object, a different plane

---

* Some of the rods have air bubbles in them, and then there results a capillary tube when they are drawn out. It is well to draw out a glass tube into a fine thread and examine it as described. The central cavity makes the experiment much more complex.

† The collodion used is a 5 per cent. solution of gun cotton in equal parts of sulphuric ether and 95 per cent. alcohol. It is well to dip the rod two or three times in the collodion and to hold it vertically while drying. The collodion will gather in drops and one will see the difference between a thick and a thin membranous covering (Fig. 25).

will be in focus and a different optical section obtained. The most satisfactory optical sections are obtained with high objectives having large aperture (§ 18).

Nearly all the transparent objects studied may be viewed in optical section. A striking example will be found in studying mammalian red blood-corpuscles on edge. The experiments with the solid glass rods (§§ 89, 91) furnish excellent and striking examples of optical sections (Pl. III, Fig. 24–26).

§ 93. **Currents in Liquids.**—Employ the 18 mm. (¾ in.) objective, and as object put a few particles of carmine on the middle of a slide, and add a drop of water. Grind the carmine well with a scalpel blade, and then cover it. If the microscope is inclined, a current will be produced in the water, and the particles of carmine will be carried along by it. Note that the particles seem to flow up instead of down, why is this (§§ 3, 34)?

Lamp-black rubbed in water containing a little mucilage answers well for this experiment.

§ 94. **Pedesis or Brownian Movement.**—Employ the same object as above, but a 5 mm. (⅕ in.) or higher objective in place of the 18 mm. Make the body of the microscope vertical, so that there may be no currents produced. Use a small diaphragm and light the field well. Focus, and there will be seen in the field large motionless masses, and between them small masses in constant motion. This is an indefinite dancing or oscillating motion.

This indefinite but continuous motion of small particles in a liquid is called *Pedesis or Brownian movement.* Also, but improperly, molecular movement, from the smallness of the particles.

The motion is increased by adding a little gum arabic solution or a slight amount of silicate of soda or of soap; sulphuric acid and various saline compounds retard or check the motion. One of the best objects is pumice stone ground finely. In this the movement is so active that it is difficult to follow the course of single particles. Pedesis is exhibited by all solid matter if finely enough divided and in a suitable liquid. No adequate explanation of this phenomenon has yet been offered. See Carpenter 182–183, Beale 195, Jevons in Quart. Jour. Science, new series, Vol. VIII, (1878), p. 167.

Compare the pedetic motion with that of a current by slightly inclining the body of the microscope. The small particles will continue their independent leaping movements while they are carried along by the current.

§ 95. **Demonstration of Pedesis with the Polarizing Microscope.**—The following demonstration shows conclusively that the pedetic motion is real and not illusive. (Ranvier, p. 173).

INTERPRETATION OF APPEARANCES.

Open the abdomen of a dead frog (an alcoholic specimen will do); turn the viscera to one side and observe the small whitish masses at the emergence of the spinal nerves. With fine forceps remove one of these and place it on the middle of a clean slide. Add a drop of water, or of water containing a little gum arabic. Rub the white mass around in the drop of liquid and soon the liquid will have a milky appearance. Remove the white mass, place a cover-glass on the milky liquid and seal the cover by painting a ring of castor oil all around it, half the ring being on the slide and half on the cover-glass. This is to avoid the production of currents by evaporation.

Put the preparation under the miroscope and examine with first a low then a higher power (3 mm. or ⅙ in.). In the field will be seen multitudes of crystals of carbonate of lime, the larger crystals are motionless but the smallest ones exhibit marked pedetic movement.

Use the micro-polariscope (see Ch. IV), light with great care and exclude all adventitious light from the microscope by shading the object (§ 66) and also by shading the eye. Focus sharply and observe the pedetic motion of the small particles, then cross the polarizer and analyzer, that is, turn one or the other until the field is dark. Part of the large motionless crystals will shine continuously and a part will remain dark, but the small crystals between the large ones will shine for an instant, then disappear, only to appear again the next instant. This demonstration is believed to furnish absolute proof that the pedetic movement is real and not illusory.

§ 96. In addition to the above experiments it is very strongly recommended that the student follow the advice of Beale, p. 248, and examine first with a low then a higher power mounted dry, then in water, lighted with reflected light, then with transmitted light, the following: Potato, wheat, rice, and corn starch, easily obtained by scraping the potato and the grains mentioned; bread crumbs; portions of feather. Portions of feather accidentally present in histological preparations have been mistaken for lymphatic vessels (B. 248). Fibers of cotton, linen and silk. Textile fibers accidentally present have been considered nerve fibers, etc. Human and animal hairs, especially cat hairs. These are very liable to be present in preparations made in this laboratory. The scales of butterflies and moths, especially the common clothes moth. The dust swept from carpeted and wood floors. Tea leaves and coffee grounds. Dust found in living rooms in places not frequently dusted. In the last will be found a regular museum of objects.

For different appearances due to the illuminator see Nelson, in Jour. Roy. Micr. Soc., 1891, pp. 90–105.

CHAPTER III.

MAGNIFICATION, MICROMETRY AND DRAWING.

APPARATUS AND MATERIAL FOR THIS CHAPTER.

Simple and compound microscope (Ch. I); Steel scale or rule divided to millimeters and ½ths; Block for magnifier and compound microscope (§ 98, 102); Dividers (§ 98, 99, 102); Stage micrometer (§ 101); Wollaston's camera lucida (§ 102, 121); Ocular micrometer (§ 112); Micrometer ocular (§ 114). Abbe camera lucida (§§ 122–127).

§ 97. **The Magnification, Amplification or Magnifying Power** of a microscope or any of its parts is the number obtained by dividing any linear dimension of the image by the corresponding linear dimension of the object magnified. For example, if the image of some object is 40 mm. long, and the actual length of the object magnified is 2 mm. the magnification is 40÷2=20.

Magnification is expressed in diameters or times linear, that is but one dimension is considered. In giving the scale at which a microscopical or histological drawing is made, the word magnification is frequently indicated by the sign of multiplication thus: ×450, upon a drawing would mean that the figure or drawing is 450 times as large as the object.

MAGNIFICATION OF A SIMPLE MICROSCOPE.

§ 98. **The Magnification of a Simple Microscope** is the ratio between the object magnified (Fig. 4, A B), and the virtual image (Fig. 4, A′ B′). To obtain the size of the image (Fig. 4, A′ B′), place the tripod magnifier near the edge of a support of such a height that the distance from the upper surface of the magnifier to the table is 250 millimeters.

As object, place a scale of some kind ruled in millimeters on the support under the magnifier. Put some white paper on the table at the base of the support, and on the side facing the light.

Close one eye, and hold the head so that the other will be near the upper surface of the lens. Focus if necessary to make the image clear (§ 4). Open the closed eye, and the image of the rule will appear as if on the paper at the base of the support. Hold the head very still, and, with dividers, get the distance between any two lines of the image. This is the so-called method of binocular or double vision in which the microscopic image is seen with one eye and the dividers with the other, the two images appearing to be fused in a single visual field.

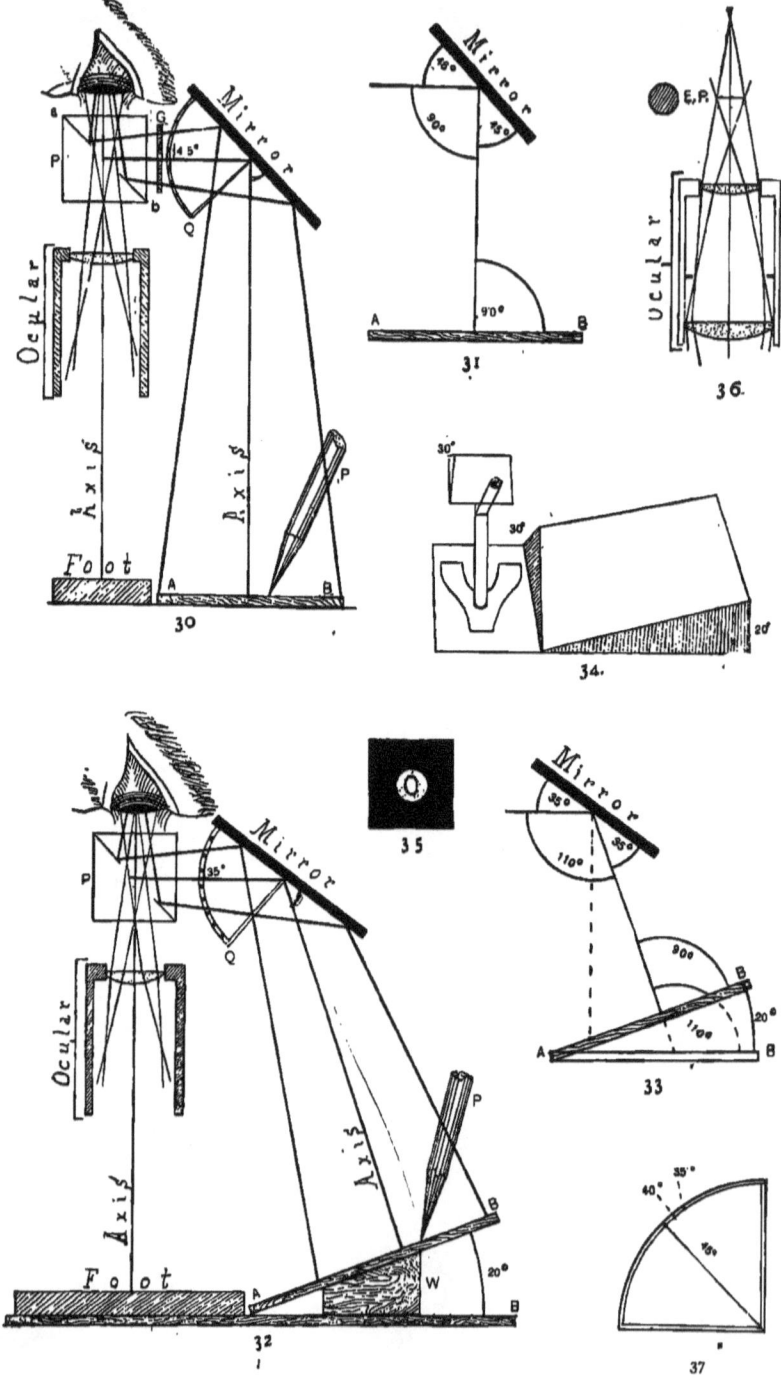

EXPLANATION OF PLATE IV.

Figures showing the use of the Abbe Camera Lucida (§§ 122–125).

Fig. 30. Abbe Camera Lucida with the mirror at 45°, the drawing surface horizontal, and the microscope vertical.

Axis, Axis. Axial ray from the microscope and from the drawing surface.
A B. Marginal rays of the field on the drawing surface. a b. Sectional view of the silvered surface in the lower of the triangular prisms composing the cubical prism (P). The silvered surface is shown as incomplete in the center, thus giving passage to the rays from the microscope.

Foot. Foot or base of the microscope.

G. Smoked glass seen in section. It is placed between the mirror and the prism to reduce the light from the drawing surface.

Mirror. The mirror of the camera lucida. A quadrant (Q) has been added to indicate the angle of inclination of the mirror, which in this case is 45°.

Ocular. Ocular of the microscope over which the prism of the camera lucida is placed.

P, P. Drawing pencil and the cubical prism over the ocular.

Fig. 31. Geometrical figure showing the angles made by the axial ray with the drawing surface and the mirror.

A B. The drawing surface.

Fig. 32. The Abbe Camera Lucida with the mirror at 35°, and the position of the drawing surface to avoid distortion (§ 124).

Axis, Axis. Axial ray from the microscope and from the drawing surface.

A B. Drawing surface raised toward the microscope 20°.

Foot. The foot or base of the microscope.

Mirror with quadrant (Q). The mirror is seen to be at an angle of 35°.

Ocular. Ocular of the microscope.

P, P. Drawing pencil, and the cubical prism over the ocular.

W. Wedge to support the drawing board.

Fig. 33. Geometrical figure of the preceding, showing the angles made by the axial ray with the mirror and the necessary elevation of the drawing board to avoid distortion. From the equality of opposite angles, the angle of the axial ray reflected at 35° must make an angle of 110° with a horizontal drawing board. The board must then be elevated toward the microscope 20° in order that the axial ray may be perpendicular to it, and thus fulfill the requirements necessary to avoid distortion (§§ 120, 124).

Fig. 34. This shows the arrangement of the drawing surface for a mirror at 35° and the microscope inclined 30° (Mrs. Gage). (§ 125).

Fig. 35. Upper view of the prism of the camera lucida. A considerable portion of the face of the prism is covered, and the opening in the silvered surface appears oval.

Fig. 36. Ocular, showing eye-point, E P. It is at this point both horizontally and vertically that the hole of the silvered surface should be placed (§ 123).

Fig. 37. Quadrant to be attached to the mirror of the Abbe Camera Lucida to indicate the angle of the mirror. As the angle is nearly always at 45°, 40° or 35°, only those angles are shown.

MAGNIFICATION AND DRAWING.

§ 99. Measuring the Spread of Dividers.—This should be done on a steel scale divided to millimeters and ½ths.

As ½ mm. cannot be see plainly by the unaided eye, place one arm of the dividers at a centimeter line, and then with the tripod magnifier count the number of spaces on the rule included between the points of the dividers. The magnifier simply makes it easy to count the spaces on the rule included between the points of the dividers—it does not, of course, increase the number of spaces or change their value.

As the distance between any two lines of the image of the scale gives the size of the virtual image (Pl. I, Fig. 4, A′ B′), and as the size of the object is known, the magnification is determined by dividing the size of the image by the size of the object. Thus, suppose the distance between the two lines of the image is measured by the dividers and found on the steel scale to be 15 millimeters, and the actual size of the space between the two lines of the object is 2 millimeters, then the magnification must be $15 \div 2 = 7\frac{1}{2}$. That is, the image is 7½ times as long or wide as the object. In this case the image is said to be magnified 7½ diameters, or 7½ times linear.

The magnification of any simple magnifier may be determined experimentally in the way described for the tripod.

MAGNIFICATION OF A COMPOUND MICROSCOPE.

§ 100. The Magnification of a Compound Microscope is the ratio between the final or virtual image (Pl. I, Fig 5, B A), and the object magnified (Pl. I, Fig. 5, $A^s B^s$).

The determination of the magnification of a compound microscope may be made as with a simple microscope (§ 98), but this is very fatiguing and unsatisfactory.

§ 101. Stage, Object or Objective Micrometer.—For determining the magnification of a compound microscope and for the purposes of micrometry it is necessary to have a finely divided scale or rule on glass or on metal. Such a finely divided scale is called a micrometer, and for ordinary work one on glass is most convenient. The spaces between the lines should be $\frac{1}{10}$ and $\frac{1}{100}$ millimeter, and when high powers are to be used the lines should be very fine. It is of advantage to have the coarser lines filled with graphite (plumbago), especially when low powers are to be used. If one has an uncovered micrometer the lines may be very readily filled by rubbing some of the plumbago on the surface with the end of a cork, the superfluous plumbago may be removed by using a clean dry cloth or a piece of the Japanese paper. After the lines are filled and the plumbago wiped from the surface, the slide

should be examined and if it is found satisfactory, *i. e.*, if the lines are black, a cover-glass on which is a drop of warm balsam may be put over the lines to protect them.

§ 102. **Determination of Magnification.**—This is most readily accomplished by the use of some form of camera lucida (§§ 121, 122), that of Wallaston being most convenient as it may be used for all powers, and the determination of the *standard distance of 250 millimeters* at which to measure the image is very readily determined (Pl. III, Fig. 27, § 104).

Employ the 18 mm. (¾ in.) objective and a 50 mm. (2 in., A or No. 1) ocular and stage micrometer as object. For this power the $\frac{1}{10}$th mm. spaces of the micrometer should be used as object. Focus sharply, and make the body of the microscope horizontal, by bending the flexible pillar, being careful not to bring any strain upon the fine adjustment (§ 71, Fig. 10).

Put a Wallaston's camera lucida (§ 121) in position, and turn the ocular around if necessary so that the broad flat surface may face directly upward as shown in Fig. 27. Elevate the microscope by putting a block under the base, so that the perpendicular distance from the upper surface of the camera lucida to the table is 250 mm. (§ 104). Place some white paper on the work-table beneath the camera lucida.

Close one eye, and hold the head so that the other may be very close to the camera lucida. Look directly down. The image will appear to be on the table. It may be necessary to readjust the focus after the camera lucida is in position. If there is difficulty in seeing dividers and image consult § 121. Measure the image with dividers and obtain the power exactly as above (§§ 98, 99).

Thus: Suppose two of the $\frac{1}{10}$th mm. spaces were taken as object, and the image is measured by the dividers, and the spread of the dividers is found on the steel rule to be $9\frac{2}{5}$ millimeters. If now the object is $\frac{2}{10}$ths of a millimeter and the magnified image is $9\frac{2}{5}$ millimeters the magnification (which is the ratio between size of object and image) must be $9\frac{2}{5} \div \frac{2}{10} = 47$. That is, the magnification is 47 diameters or 47 times linear. If the fractional numbers in the above example trouble the student, both may be reduced to the same denomination, thus: If the size of the image is found to be $9\frac{2}{5}$ mm. this number may be reduced to tenths mm. so it will be of the same denomination as the object. In 9 mm. there are 90 tenths, and in $\frac{2}{5}$ there are 4 tenths, then the whole length of the image is $90 + 4 = 94$ tenths of a millimeter. The object is 2 tenths of a millimeter, then there must have been a magnification of $94 \div 2 = 47$ diameters in order to produce an image 94 tenths of a millimeter long.

Put the 25 mm. (1 in. C or No. 4) ocular in place of one of 50 mm.

Focus, and then put the camera lucida in position. Measure the size of the image with dividers and a rule as before. The power will be considerably greater than when the low ocular was used. This is because the virtual image (Fig. 5, B′ A′), seen with the high ocular is larger than the one seen with the low one. The real image (Fig. 5, A B), remains nearly the same, and would be just the same if positive, par-focal oculars (§§ 21, 48 note), were used.

Lengthen the body of the microscope 50–60 mm. by pulling out the draw-tube. Remove the camera lucida, and focus, then replace the camera, and obtain the magnification. It will be greater than with the shorter body. This is because the real image (Fig. 5, B A) is formed farther from the objective when the body is lengthened, and being formed farther from the objective it must necessarily be larger (§ 7 and Fig. 28).

§ 103. **Varying the Magnification of a Compound Microscope.** It will be seen from the above experiments (§ 102), that independently of the distance at which the microscopic image is measured (§ 104), there are three ways of varying the power of a compound microscope. These are named below in the order of desirability.

(1) *By using a higher or lower objective.*
(2) *By using a higher or lower ocular.*
(3) *By lengthening or shortening the tube of the microscope.**

§ 104. **Standard Distance of 250 Millimeters at which the Virtual Image is Measured.**—For obtaining the magnification of both the simple and the compound microscope the directions were to measure the virtual image at a distance of 250 millimeters. This is not that the image could not be seen and measured at any other distance, but because some standard must be selected, and this is the most common one. The necessity for the adoption of some common standard will be seen at a glance in Pl. III, Fig. 28, where is represented graphically the fact that the size of the virtual image depends directly on the distance at which it is projected, and this size is directly proportional to the vertical distance from the apex of the triangle, of which it forms a base. The distance of 250 millimeters has been chosen on the supposition that it is the distance of most distinct vision for the normal human eye.

---

* **Amplifier.**—In addition to the methods of varying the magnification given in ₴ 103, the magnification is sometimes increased by the use of an amplifier, that is a diverging lens or combination placed between the objective and ocular and serving to give the image forming rays from the objective an increased divergence. This accessory was first made effective by Tolles, who made it in the form of a small achromatic concavo-convex lens to be screwed into the lower end of the draw-tube (Pl. II, Fig. 10) and thus but a short distance above the objective. The divergence given the rays increases the size of the real image about two fold.

Demonstrate the difference in magnification due to the distance at which the image is projected, by raising the microscope so that the distance will be 350 millimeters, then 150 millimeters.

In preparing drawings it is often of great convenience to make them at a distance somewhat less or somewhat greater than the standard. In such a case the magnification must be determined for the special distance.

It may be remarked further that if spectacles are not used, a near-sighted (myopic) person would obtain a somewhat greater, and a far-sighted (presbyopic) person a somewhat less magnification for the same instrument at the standard distance. This is because the eye of the observer forms an integral optical part of the microscope at the time of observation, and the equivalent focus of the myopic eye is less than normal and that of the presbyopic eye greater (§ 7).

For discussions of the magnification of the microscope, see: B., pp. 41, 355; C., pp. 161, 206; N. & S., p. 176; R., p. 29; Robin, p. 126; Amer. Soc. Micrs., 1884, p. 183; 1889, p. 22; Amer. Jour. Arts and Sciences, 1890, p. 50; Jour. Roy. Micr. Soc., 1888, 1889.

§ 105. **Table of Magnifications and of the Valuations of the Ocular Micrometer.**—*The following table should be filled out by each student. In using it for Micrometry and Drawing it is necessary to keep clearly in mind the exact conditions under which the determinations were made, and also the ways in which variation in magnification and the valuation of the ocular micrometer may be produced* (§§ 103, 104, 114, 116).

OBJECT-IVE.	OCULAR 50 mm.		OCULAR 25 mm.		OCULAR MICROMETER VALUATION.	
	TUBE IN.	TUBE OUT 50 MM.	TUBE IN.	TUBE OUT —MM.	TUBE IN.	OUT —MM.
¾, 18 mm.	×42	×56	×58	×84	50 mm	85 mm
⅙, 3 mm.	×	×	×	×		
¼	×118	×164	×224	×290	180 m	230 mm
SIMPLE MICROSCOPE.	×10					

MICROMETRY.

§ 106. **Micrometry** is the determination of the size of objects by the aid of a microscope.

MICROMETRY WITH THE SIMPLE MICROSCOPE.

§ 107. With a simple microscope, (A) the easiest and best way is to use dividers and then the simple microscope to see when the points of the dividers exactly include the object. The spread of the dividers is then obtained as above (§ 99). This amount will be the actual size of the object, as the microscope was only used in helping to see when the divider points exactly enclosed the object, and then for reading the divisions on the rule in getting the spread of the dividers.

(B) One may put the object under the simple microscope and then as determining the power (§ 98), measure the image at the standard distance. If now the size of the image so measured is divided by the magnification of the simple microscope, the quotient will give the actual size of the object.

Use a fly's wing or some other object of about that size and try to determine the width in the two ways described above. If all the work is accurately done the results will agree.

MICROMETRY WITH THE COMPOUND MICROSCOPE.

There are several ways of varying excellence for obtaining the size of objects with the compound microscope, the method with the ocular micrometer (§§ 116, 117) being most accurate.

§ 108. **Unit of Measure in Micrometry.**—As most of the objects measured with the compound microscope are smaller than any of the originally named divisions of the meter, and the common or decimal fractions necessary to express the size are liable to be unnecessarily cumbersome, *Harting*, in his work on the microscope (1859), proposed the one thousandth of a millimeter ($\frac{1}{1000}$ mm. or 0.001 mm.) or one millionth of a meter ($\frac{1}{1000000}$ or 0.000001 meter) as the unit. He named this unit micro-millimeter and designated it mmm. In 1869, *Listing* (Carl's Repetorium für Experimental-Physik, Bd. X, P. 5) favored the thousandth of a millimeter as unit and introduced the name **Mikron** or *micrum*. In English it is most often written *Micron*, plural *micra* or *microns*, pronunciation Mĭc'rŏn, or Mī'crŏn. By universal consent the sign or abbreviation used to designate it is the Greek μ. Adopting this unit and sign, one would express five thousandths of a millimeter ($\frac{5}{1000}$ or 0.005ths mm.) thus, 5μ.*

---

*The term Micromillimeter ab. mmm. is very cumbersome, and besides is entirely inappropriate since the adoption of definite meanings for the prefixes micro and mega, meaning respectively one millionth and one million times the unit before which it is placed. A micromillimeter would then mean one-millionth of a millimeter, not one-thousandth. The term micron, has been adopted by the great microscopical societies, the international commission on weights and measures and by original investigators, and is in the opinion of the writer the best term to employ. Jour. Roy. Micr. Soc., 1888, p. 502; Nature, Vol. XXXVII, (1888), p. 388.

§ 109. **Micrometry** *by the use of a stage micrometer on which to mount the object.*—In this method the object is mounted on a micrometer and then put under the microscope and the number of spaces covered by the object is read off directly. It is exactly like putting any large object on a rule and seeing how many spaces of the rule it covers. The defect in the method is that it is impossible to properly arrange objects on the micrometer. Unless the objects are circular in outline they are liable to be oblique in position and in every case the end or edges of the object may be in the middle of a space instead of against one of the lines, consequently the size must be estimated or guessed at rather than really measured.

§ 110. **Micrometry** *by dividing the size of the image by the magnification of the microscope.*—For example, employ the 3 mm. objective, 25 mm. ocular, and a Necturus' red blood-corpuscle preparation as object.* Obtain the size of the image of the long and short axes of three corpuscles with the camera lucida and dividers exactly as in obtaining the magnification of the microscope (§ 102). Divide the size of the image in each case by the magnification and the result will be the actual size of the blood-corpuscle. Thus, suppose the image of the long axis of the corpuscle is 18 mm. and the magnification of the microscope 400 diameters (§ 97), then the actual length of this long axis of the corpuscle is 18 mm. ÷ 400 = .045 mm. or 45 μ (§ 108).

§ 111. **Micrometry** *by the use of a Stage Micrometer and a Camera Lucida.*—Employ the same object, objective and ocular as before. Put the camera lucida in position, and with a lead pencil make dots on the paper at the limits of the image of the blood-corpuscle. Measure the same three that were measured in § 110.

Remove the object, place the stage micrometer under the microscope, focus well, and draw the lines of the stage micrometer so as to include the dots representing the limits of the part of the image to be measured. As the value of the spaces on the stage micrometer is known, the size of the object is determined by the number of spaces of the micrometer required to include it.

This simply enables one to put the image of a fine rule on the image of a microscopic object. It is theoretically an excellent method, and nearly the same as measuring the spread of the dividers with a simple microscope (§§ 99, 117).

OCULAR MICROMETER.

§ 112. **Ocular Micrometer, Eye-Piece Micrometer.**—This, as the name implies, is a micrometer to be used with the ocular. It is a

---

* As the same three blood corpuscles are to be measured in three ways, it is an advantage to put a delicate ring around a group of three or more corpuscles and make a sketch of the whole enclosed group, marking on the sketch the corpuscles measured. The different corpuscles vary considerably in size, so that accurate comparison of different methods of measurement can only be made when the same corpuscles are measured in each of the ways.

micrometer on glass, and the lines are sufficiently coarse to be clearly seen by the ocular. The lines should be equidistant and about $\frac{1}{10}$th or $\frac{1}{20}$th mm. apart and every fifth line should be longer and heavier to facilitate counting. If the micrometer is ruled in squares (*net-micrometer*) it will be very convenient for many purposes.

The ocular micrometer is placed in the ocular, no matter what the form of the ocular (*i. e.*, whether positive or negative), at the level at which the real image is formed by the objective, and the image appears to be immediately upon or under the ocular micrometer and hence the number of spaces on the ocular micrometer required to measure the real image may be read off directly. This is measuring the size of the real image, however, and the actual size of the object can only be determined by determining the ratio between the size of the real image and the object. In other words it is necessary to get the *valuation of the ocular micrometer* in terms of a stage micrometer.

§ 113. **Valuation of the Ocular Micrometer.**—This is the value of the divisions of the ocular micrometer for the purposes of micrometry, and is entirely relative, depending upon the magnification of the real image formed by the objective, consequently it changes with every change in the magnification of the real image and must be specially determined for every optical combination (*i. e.*, objective and ocular) and for every change in the length of the tube of the microscope. That is, it is necessary to determine the ocular micrometer valuation for every condition modifying the real image of the microscope (§ 103).

§ 114. **Obtaining the Ocular Micrometer Valuation.**—As an example, employ the 25 mm. ocular and 18 mm. objective. Place the stage micrometer under the microscope for an object, and put the ocular micrometer in position, either through a slit in the ocular, or remove the eye-lens and place it upon the ocular diaphragm.*

Light the field well, and look into the microscope. The lines on the ocular micrometer should be very sharply defined. If they are not, raise or lower the eye-lens to make them so ; that is, focus as with the simple magnifier.

When the lines of the ocular micrometer are distinct, focus the mi-

---

* It is a great convenience to have a micrometer ocular (§ 25) with a spring and screw to enable one to accurately place the ocular micrometer. Any negative ocular may, however, be used as a micrometer ocular by placing the ocular micrometer at the level of the ocular diaphragm, that is where the real image is formed. This is very conveniently arranged for by some opticians by a slit in the side of the ocular, and the ocular micrometer is mounted in some way and simply introduced through the opening in the side. When no side opening exists the mounting of the ocular may be unscrewed and the ocular micrometer, if on a cover-glass, can be laid on the upper side of the ocular diaphragm.

croscope (§§ 32, 37) for the stage micrometer. The image of the stage micrometer will appear to be directly under or upon the ocular micrometer.

Make the lines of the two micrometers parallel by rotating the ocular, or changing the position of the stage micrometer, or both if necessary, and then make any two lines of the stage micrometer coincide with any two on the ocular micrometer. To do this it may be necessary to pull out the draw-tube a greater or less distance. See how many spaces are included on each of the micrometers.

Divide the value of the included space or spaces on the stage micrometer by the number of divisions on the ocular micrometer required to include them, and the quotient so obtained will give the valuation of the ocular micrometer in fractions of the unit of measure of the stage micrometer. For example, suppose the millimeter is taken as the unit for the stage micrometer and this unit is divided into spaces of $\frac{1}{10}$th and $\frac{1}{100}$th millimeter. If now, with a given optical combination and tube-length, it requires 10 spaces on the ocular micrometer to include the real image of $\frac{1}{10}$th millimeter on the stage micrometer, obviously one space on the ocular micrometer would include only one-tenth as much, or $\frac{1}{10}$th mm. ÷ 10 = $\frac{1}{100}$th mm. That is, each space on the ocular micrometer would include $\frac{1}{100}$th of a millimeter on the stage micrometer, or $\frac{1}{100}$th millimeter of length of any object under the microscope, the conditions remaining the same. Or in other words, it would require 100 spaces on the ocular micrometer to include 1 millimeter on the stage micrometer, then as before 1 space of the ocular micrometer would have a valuation of $\frac{1}{100}$th millimeter for the purposes of micrometry; and the size of any minute object may be determined by multiplying this valuation of one space by the number of spaces required to include it. For example, suppose the fly's wing or some part of it covered 8 spaces on the ocular micrometer, it would be known that the real size of the part measured is $\frac{1}{100}$th mm. × 8 = $\frac{8}{100}$th or 80 μ (§ 108).

§ 115. **Varying the Ocular Micrometer Valuation.**—Any change in the objective, the ocular or the tube-length of the microscope, that is to say any change in the size of the real image, produces a corresponding change in the ocular micrometer valuation (§ 103, 112).

§ 116. **Micrometry with the Ocular Micrometer.**—Use the 3 mm. objective and preparation of Necturus blood corpuscles as object. Make certain that the tube of the microscope is of the same length as when determining the ocular micrometer valuation. In a word be sure that all the conditions are exactly as when the valuation was determined, then put the preparation under the microscope and find the same three red corpuscles that were measured in the other ways (§§ 110, 111).

Count the divisions on the ocular micrometer required to enclose or measure the long and the short axis of each of the three corpuscles, then multiply the number of spaces in each case by the valuation of the ocular micrometer for this objective, tube-length and ocular, and the results will give the actual length of the axes of the corpuscles in each case.

The same corpuscle is, of course, of the same actual size, when measured in each of the three ways (§§ 110, 111, 116), so that if the methods are correct and the work carefully enough done the same results should be obtained by each method. See general remarks on micrometry (§ 117).*

---

* There are three ways of using the ocular micrometer, or of arriving at the size of the objects measured with it:

(A) By finding the value of a division of the ocular micrometer for each optical combination and tube-length used, and employing this valuation as a multiplier. This is the method given in the text and is the one most frequently employed. Thus, suppose with a given optical combination and tube-length it required five divisions on the ocular micrometer to include the image of $\frac{2}{10}$ths millimeter of the stage micrometer, then obviously one space on the ocular micrometer would include $\frac{1}{5}$th of $\frac{2}{10}$ths mm. or $\frac{1}{25}$th mm.; and the size of any unknown object under the microscope would be obtained by multiplying the number of divisions on the ocular micrometer required to include its image by the value of one space, or in this case, $\frac{1}{25}$th mm. Suppose some object, as the fly's wing required 15 spaces of the ocular micrometer to include some part of it, then the actual size of this part of the wing would be $15 \times \frac{1}{25} = \frac{3}{5}$ths, or 0.6 mm.

(B) By finding the number of divisions on the ocular micrometer required to include the image of an entire millimeter of the stage micrometer, and using this number as a divisor. This number is also sometimes called the *ocular micrometer ratio*. Taking the same case as in (A) suppose five divisions of the ocular micrometer are required to include the image of $\frac{2}{10}$ths mm., on the stage micrometer, then evidently it would require $5 \div \frac{2}{10} = 25$ divisions on the ocular micrometer to include a whole millimeter on the stage micrometer, then the number of divisions of the ocular micrometer required to measure an object divided by 25 would give the actual size of the object in millimeters or in a fraction of a millimeter. Thus, suppose it required 15 divisions of the ocular micrometer to include the image of some part of the fly's wing, the actual size of the part included would be $15 \div 25 = \frac{3}{5}$ or 0.6 mm. This method is really exactly like the one in (A), for dividing by 25 is the same as multiplying by $\frac{1}{25}$th.

(C). By having the ocular micrometer ruled in millimeters and divisions of a millimeter, and then getting the size of the real image in millimeters. In employing this method a stage micrometer is used as object and the size of the image of one or more divisions is measured by the ocular micrometer, thus: Suppose the stage micrometer is ruled in $\frac{1}{10}$th and $\frac{1}{100}$th mm. and the ocular micrometer is ruled in millimeters and $\frac{1}{10}$th mm. Taking $\frac{1}{10}$th mm. on the stage micrometer as object, as in the other cases, suppose it requires 10 of the $\frac{1}{10}$th mm. spaces or 1 mm. to measure the real image, then the real image must be magnified $\frac{10}{10} \div \frac{2}{10} = 5$ diameters, that is the real image is five times as great in length as the object, and the size of an object may be determined by putting it under the micro-

§ 117. **Remarks on Micrometry.**—In using adjustable objectives (§§ 16, 63), the magnification of the objective varies with the position of the adjusting collar, being greater when the adjustment is closed as for thick cover-glasses than when open, as for thin ones. This variation in the magnification of the objective produces a corresponding change in the magnification of the entire microscope and the ocular micrometer valuation—therefore it is necessary to determine the magnification and ocular micrometer valuation for each position of the adjusting collar.

While the principles of micrometry are simple, it is very difficult to get the exact size of microscopic objects. This is due to the lack of perfection and uniformity of micrometers, and the difficulty in determining the exact limits of the object to be measured. Hence, all microscopic measurements are only approximately correct, the error lessening with the increasing perfection of the apparatus and the skill of the observer.

A difficulty when one is using high powers is the width of the lines of the micrometer. If the micrometer is perfectly accurate half the width of each line belongs to the contiguous spaces, hence one should measure the image of the space from the centers of the lines bordering the space, or as this is somewhat difficult in using the ocular micrometer, one may measure from the inside of one bordering line and from the outside of the other. If the lines are of equal width this is as accurate as measuring from the center of the lines. Evidently it would not be right to measure from either the inside or the outside of both lines (Pl. III, Fig. 29).

It is also necessary in micrometry, to use an objective of sufficient power to enable one to see all the details of an object with great distinctness. The necessity of using sufficient amplification in micrometry has been especially remarked upon by Richardson, Monthly Micr. Jour., 1874, 1875; Rogers, Proc. Amer. Soc. Microscopists, 1882, p. 239; Ewell, North American Pract., 1890, pp. 97, 173.

As to the limit of accuracy in micrometry, one who has justly earned the right to speak with authority expresses himself as follows: "*I assume that o 2µ is the limit of precision in microscopic measures, beyond which it is impossible to go with certainty.*" W. A. Rogers, Proc. Amer. Soc. Micrs., 1883, p. 198.

scope and getting the size of the real image in millimeters with the ocular micrometer and dividing it by the magnification of the real image, which in this case is 5 diameters.

Use the fly's wing as object as in the other cases, and measure the image of the same part. Suppose that it required 30 of the $\frac{1}{10}$ mm. divisions = $\frac{30}{10}$ mm. or 3 mm. to include the image of the part measured, then evidently the actual size of the part measured would be 3 mm. ÷ 5 = $\frac{3}{5}$ mm., the same result as in the other cases.

.. In comparing these methods it will be seen that in the first two (A and B) the ocular micrometer may be simply ruled with equidistant lines without regard to the absolute size in millimeters or inches of the spaces. In the last method the ocular micrometer must have its spaces some known division of a millimeter or inch. In the first two methods only one standard of measure is required, viz., the stage micrometer; in the last methods two standards must be used,—a stage micrometer and an ocular micrometer. Of course the ocular micrometer in the first two cases must have the lines equidistant as well as in the last case, but ruling lines equidistant and an exact division of a millimeter or an inch are two quite different matters.

MAGNIFICATION AND DRAWING.

In comparing the methods of micrometry with the compound microscope, given above (§§ 109, 110, 111, 116) the one given in § 109 is impracticable, that given in § 110 is open to the objection that two standards are required,—the stage micrometer, and the steel rule; it is open to the further objection that several different operations are necessary, each operation adding to the probability of error. Theoretically the method given in § 111 is good, but it is open to the very serious objection in practice that it requires so many operations which are especially liable to introduce errors. The method that experience has found most safe and expeditious, and applicable to all objects, is the method with the ocular micrometer. If the valuation of the ocular micrometer has been accurately determined, then the only difficulty is in deciding on the exact limits of the object to be measured and so arranging the ocular micrometer that these limits are inclosed by some divisions of the micrometer. Where the object is not exactly included by whole spaces on the ocular micrometer, the chance of error comes in, in estimating just how far into a space the object reaches on the side not in contact with one of the micrometer lines. If the ocular micrometer has some quite narrow spaces, and others considerably larger, one can nearly always manage to exactly include the object by some two lines.

For those especially interested in micrometry, as in its relation to medical jurisprudence, the following references are recommended. These articles consider the problem in a scientific as well as a practical spirit: The papers of Prof. Wm. A. Rogers on micrometers and micrometry in the Amer. Quar. Micr. Jour., Vol. I, pp. 97, 208; Proceedings Amer. Soc. Microscopists, 1882, 1883, 1887. Dr. M. D. Ewell, Proc. Amer. Soc. Micrs., 1890; The Microscope, 1889, pp. 43–45; North Amer. Pract., 1890, pp. 97, 173. Dr. J. J. Woodward, Amer Jour. of the Med. Sci., 1875. M. C. White, Article Blood-stains, Ref. Hand-Book, Med. Sciences, 1885. For the change in magnification due to a change in the adjustment of adjustable objectives, see Jour. Roy. Micr. Soc., 1880, p. 702; Amer. Monthly Micr. Jour., 1880, p. 67.

DRAWING WITH THE MICROSCOPE.

§ 118. Microscopic objects may be drawn free-hand directly from the microscope, but in this way a picture giving only the general appearance and relations of parts is obtained. For pictures which shall have all the parts of the object in true proportions and relations, it is necessary to obtain an exact outline of the image of the object, and to locate in this outline all the principal details of structure. It is then possible to complete the picture free hand from the appearance of the object under microscope. The appliance used in obtaining outlines, etc., of the microscopic image is known as a *camera lucida*.

§ 119. Camera Lucida.—This is an optical apparatus for enabling one to see objects in greatly different situations, as if in one field of vision, and with the same eye. In other words, it is an optical device for superimposing or combining two fields of view in one eye.

As applied to the microscope, it causes the magnified virtual image of the object under the microscope to appear as if projected upon the table or drawing board, where it is visible with the drawing paper, pencils, dividers, etc., by the same eye, and in the same field of vision. The microscopic image appears like a picture on the drawing paper. This is accomplished in two distinct ways:

(A) By a camera lucida reflecting the rays from the microscope so that their direction when they reach the eye coincides with that of the rays from the drawing

paper, pencils, etc. In some of the camera lucidas of this group (Wollaston's Pl. III, Fig. 27), the rays are reflected twice and the image appears as when looking directly into the microscope. In others, the rays are reflected but once and the image has the inversion produced by a plane mirror. For drawing purposes this inversion is a great objection, as it is necessary to similarly invert all the details added free-hand.

(B) By a camera lucida reflecting the rays of light from the drawing paper, etc., so that their direction when they reach the eye coincides with the direction of the rays from the microscope (Pl. IV, Fig. 30, 32). In all of the camera lucidas of this group, the rays from the paper are twice reflected and no inversion appears.

The better forms of camera lucidas (Wollaston's, Grunow's, Abbe's, etc.), may be used for drawing both with low and with high powers. Some require the microscope to be inclined (Fig. 27), while others are designed to be used on the microscope in a vertical position. As in biological work it is often necessary to have the microscope vertical, this form is to be preferred. [(B. 31, 355), (Beh. 110), (C. 112), Car. 73), (J. R. M. S. nearly every volume). For drawing at a magnification of from 5 to 100 diameters for large objects see (W. 132), (J. R. M. S. 1881, p. 819; 1882, p. 402; 1884, p. 115; 1888, p. 113), (Amer. Naturalist, 1886, p. 1071; 1887, pp. 1040, 1043)].

§ 120. **Avoidance of Distortion.**—*In order that the picture drawn by the aid of a camera lucida may not be distorted, it is necessary that the axial ray from the image on the drawing surface shall be at right angles to the drawing surface* (Pl. III, Fig. 27, Pl. IV).

§ 121. **Wollaston's Camera Lucida.**—This is a quadrangular prism of glass put in the path of the rays from the microscope, and serves to change the direction of the axial ray 90 degrees. In using it the microscope is made horizontal, and the rays from the microscope enter one half of the pupil while rays from the drawing surface enter the other half of the pupil. As seen in the figure (Pl. III, Fig. 27) the fields party overlap, and where they do so overlap, pencil or dividers and microscopic image can be seen together.

In drawing or using the dividers with the Wollaston camera lucida it is necessary to have the field of the microscope and the drawing surface about equally lighted. If the drawing surface is too brilliantly lighted the pencil or dividers may be seen very clearly but the microscopic image will be obscure. On the other hand, if the field of the microscope has too much light the microscopic image will be very definite, but the pencil or dividers will not be visible. Again, as rays from the microscope and from the drawing surface must enter independent parts of the pupil of the same eye, one must hold the eye so that the pupil is partly over the camera lucida and partly over the drawing surface. One can tell the proper position by trial. This is not a very satisfactory camera to draw with, but it is a very good form to measure the vertical distance of 250 mm. at which the drawing surface should be placed when determining magnification (§ 104).

§ 122. **Abbe Camera Lucida.**—This consists of a cube of glass cut into two triangular prisms and silvered on the lower one. A small oval hole is then cut out of the center of the silvered surface and the two prisms are cemented together, thus giving a cubical prism with a perforated 45-degree mirror (Pl. IV, Fig. 30, a b). The upper surface of the prism is covered by a perforated metal plate (Fig. 35). This prism is placed over the ocular in such a way that the light from the microscope

passes through the hole in the silvered face and thence directly to the eye. Light from the drawing surface is reflected by a mirror to the silvered surface of the prism and reflected by this surface to the eye in company with the rays from the microscope, so that the two fields appear as one, and the image is seen as if on the drawing surface (Fig. 30). It is designed for use with a vertical microscope, but see § 124.

§ 123. **Arrangement of the Camera Lucida Prism.**—In placing this camera lucida over the ocular for drawing or the determination of magnification, the center of the hole in the silvered surface is placed in the optic axis of the microscope. This is done by properly arranging the centering screws that clamp the camera to the microscope tube or ocular. The perforation in the silvered surface must also be at the level of the eye-point (Fig. 36). In other words, the prism must be so arranged vertically and horizontally that the hole in the silvered surface will be co-incident with the eye-point of the ocular. If it is above or below or to one side, part or all of the field of the microscope will be cut off. As stated above, the centering screws are for the proper horizontal arrangement of the prism. The prism is set at the right height by the makers for the eye-point of a medium ocular. If one desires to use an ocular with the eye-point farther away or nearer, as in using high or low oculars (§ 36), the position of the eye-point may be determined as directed in § 36 and the prism loosened and raised or lowered to the proper level; but in doing this one should avoid setting the prism obliquely to the mirror.

One can determine when the camera is in a proper position by looking into the microscope through it. If the field of the microscope appears as a circle and of about the same size as without the camera lucida, then the prism is in a proper position. If one side of the field is dark, then the prism is to one side of the center; if the field is considerably smaller than when the prism is turned off the ocular (§ 124) it indicates that it is not at the correct level, *i. e.*, it is above or below the eye-point.

§ 124. **Arrangement of the Mirror and the Drawing Surface.**— The Abbe camera lucida was designed for use with a vertical microscope (Fig. 30). On a vertical microscope, if the mirror is set at an angle of 45° the axial ray will be at right angles with the table top or a drawing board which is horizontal (Fig. 30), and a drawing made under these conditions would be in true proportion and not distorted. The stage of most microscopes, however, extends out so far at the sides that with a 45° mirror the image appears in part on the stage of the microscope. In order to avoid this the mirror may be depressed to some point below 45°, say at 40° or 35° (Fig. 32, 33). But as the axial ray from the mirror to the prism must still be reflected horizontally, it fol-

lows that the axial ray will no longer form an angle of 90 degrees with the drawing surface, but a greater angle. If the mirror is depressed to 35°, then the axial ray must make an angle of 110° with a horizontal drawing surface, see the geometrical figure, (Fig. 33). To make the angle 90° again so that there shall be no distortion, the drawing board must be raised toward the microscope 20°. **The general rule is to raise the drawing board twice as many degrees toward the microscope as the mirror is depressed below 45°.** Practically the field for drawing can always be made free of the stage of the microscope, at 45°, at 40° or at 35°. In the first case (45° mirror) the drawing surface should be horizontal, in the second case (40° mirror) the drawing surface should be elevated 10°, and in the third case (35° mirror) the drawing board should be elevated 20° toward the microscope. Furthermore it is necessary in using an elevated drawing board to have the mirror bar project directly laterally so that the edges of the mirror will be in planes parallel with the edges of the drawing board, otherwise there will be front to back distortion, although the elevation of the drawing board would avoid right to left distortion. If one has a micrometer ruled in squares (*net-micrometer*) the distortion produced by not having the axial ray at right angles with the drawing surface may be very strikingly shown. For example, set the mirror at 35° and use a horizontal drawing board. With a pencil make dots at the corners of some of the squares, and then with a straight edge connect the dots. The figures will be considerably longer from right to left than from front to back. Circles in the object would appear as ellipses in the drawings, the major axis being from right to left.

The angle of the mirror may be determined with a protractor, but that is troublesome. It is much more satisfactory to have a quadrant attached to the mirror and an indicator on the projecting arm of the mirror. If the quadrant is graduated throughout its entire extent, or preferably at three points, 45°, 40° and 35°, one can set the mirror at a known angle in a moment, then the drawing board can be hinged and the elevation of 10° and 20° determined with a protractor. The drawing board is very conveniently held up by a broad wedge. By marking the |position of the wedge for 10° and 20° the protractor need be used but once, then the wedge may be put into position at any time for the proper elevation.

§ 125. **Abbe Camera and Inclined Microscope.**—It is very fatiguing to draw continuously with a vertical microscope, and many mounted objects admit of an inclination of the microscope, when one can sit and work in a more comfortable position. The Abbe camera is as perfectly adapted to use with an inclined as with a vertical microscope. All that is requisite is to be sure that the fundamental law is

observed regarding the axial ray of the image and the drawing surface, viz., that they should be at right angles (Pl. IV). This is very easily accomplished as follows: The drawing board is raised toward the microscope twice as many degrees as the mirror is depressed below 45° (§ 124) then it is raised exactly as many degrees as the microscope is inclined and in the same direction, that is so the end of the drawing board shall be in a plane parallel with the stage of the microscope. The mirror must have its edges in planes parallel with the edges of the drawing board also (Fig. 34).*

§ 126. **Drawing with the Abbe Camera Lucida.**—(A) The light from the microscope and from the drawing surface should be of nearly equal intensity so that the image and the drawing pencil can be seen with about equal distinctness. This may be accomplished with very low powers (18 mm. and lower objectives) by covering the mirror with white paper when transparent objects are to be drawn. When opaque objects, like dark insects, are to be drawn it may be necessary to concentrate light upon them with a condensing lens or concave mirror. For high powers it is best to use an Abbe illuminator. Often the light may be balanced by using a larger or smaller opening in the diaphragm. One can tell which field is excessively illuminated, for it is the one in which objects are most distinctly seen. If it is the microscopic, then the image of the microscopic object is very distinct and the pencil is invisible or very indistinct. If the drawing surface is too brilliantly lighted the pencil can be seen clearly, but the microscopic image will be very obscure.

If the drawing surface is too brilliantly illuminated, it may be shaded by placing a book or a ground glass screen between it and the window, also by putting one or more smoked glasses in the path of the rays from the mirror (Fig. 30, G). If the light in the microscope is too intense, it may be lessened by using the white paper over the mirror, or by a ground glass screen between the microscope mirror and the source of light (Piersol, Amer. M. M. Jour., 1888, p. 103). It is also an excellent plan to blacken the end of the drawing pencil with carbon ink. Sometimes it is easier to draw on a black surface, using a white pencil or style. The carbon paper used in manifolding letters, etc., may be used, or ordinary black paper may be lightly rubbed on one side with a moderately soft lead pencil. Place the black paper over white paper and trace the outlines with a pointed style of ivory or bone. A corresponding dark line will appear on the white paper beneath. (J. R. M. S., 1883, p. 423).

(A) It is desirable to have the drawing paper fastened with thumb

---

*This method of using the Abbe camera lucida on an inclined microscope was devised by Mrs. Gage.

tacks, or in some other way. (B) The lines made while using the camera lucida should be very light, as they are liable to be irregular. (C) Only outlines are drawn and parts located with a camera lucida. Details are put in free-hand. (D) It is sometimes desirable to draw the outline of an object with a moderate power and add the details with a higher power. If this is done it should always be clearly stated. It is advisable to do this only with objects in which the same structure is many times duplicated, as a nerve or a muscle. In such an object all the different structures could be shown, and by omitting some of the fibers the others could be made plainer without an undesirable enlargement of the entire figure.

(E) If a drawing of a given size is desired and it cannot be obtained by any combination of oculars, objectives and lengths of the body of the microscope, the distance between the camera lucida and the table may be increased or diminished until the image is of the desired size. The image of a few spaces of the micrometer, will give the scale of enlargement, or the power may be determined for the special case (§ 127).

(F) It is of the greatest advantage, as suggested by Heinsius (Zeit. w. Mikr., 1889, p. 367), to have the camera lucida hinged so that the prism may be turned off the ocular for a moment's glance at the preparation, and then returned in place without the necessity of loosening screws and readjusting the camera. This form is now made by Zeiss, but as yet no quadrant is added. Any skilled mechanic can add that, however.

[(B. 31, 355), (Beh. 110), (C. 112), (Fol. 70), (Frey 38), (J. R. M. S. 1883, pp. 283, 560; 1886, 516; 1888, pp. 113, 809, 798); (Amer. M. M. Jour., 1888, p. 103; 1890, p. 94); (Z. w. M. 1884, p. 1-21; 1889, p. 367)].

§ 127. **Magnification of the Microscope and Size of Drawings with the Abbe Camera Lucida.**—In determining the standard distance of 250 millimeters at which to measure the image in getting the magnification of the microscope, it is necessary to measure from the point marked P on the prism (Fig. 30) to the axis of the mirror and then vertically to the drawing board.

In getting the scale at which a drawing is enlarged the best way is to remove the preparation and put in its place a stage micrometer, and to trace a few (5 or 10) of its lines under one corner of the drawing. The value of the spaces of the micrometer being given, thus,

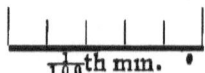

The enlargement of the figure can then be accurately determined at any time by measuring with a steel scale the length of the image of the micrometer spaces and dividing it by their known width.

Thus, suppose the 5 spaces of the scale of enlargement given with a drawing were found to measure 25 millimeters and the spaces on the micrometer were $\frac{1}{100}$th millimeter, then the enlargement would be $25 \div \frac{5}{100} = 500$. That is, the image was drawn at a magnification of 500 diameters.

If the micrometer scale is used with every drawing, there is no need of troubling one's self about the exact distance at which the drawing is made, convenience may settle that, as the special magnification in each case may be determined from the scale accompanying the picture. It should be remembered, however, that the conditions when the scale is drawn must be exactly as when the drawing was made.

CHAPTER IV.

MICRO-SPECTROSCOPE AND POLARISCOPE.

APPARATUS AND MATERIAL REQUIRED FOR THIS CHAPTER.

Compound microscope (Ch. I); Micro-spectroscope (§ 128); Watch-glasses and small vials, slides and covers (§§ 143, 144); Various substances for examination (as blood and ammonium sulphide, permanganate of potash, chlorophyll, some colored fruit, etc. (§§ 144–145); Micro-polarizer (§ 150); Selenite plate (§157 f.); Various doubly refracting objects, as crystals, textile fibers, starch, section of bone, etc.

MICRO-SPECTROSCOPE.

§ 128.—A Micro-Spectroscope, Spectroscopic, or Spectral Ocular, is a direct vision spectroscope in connection with a microscopic ocular. The one devised by Abbe and made by Zeiss consists of a direct vision spectroscope prism of the Amici pattern, and of considerable dispersion, placed over the ocular of the microscope. This direct vision or Amici prism consists of a single triangular prism of heavy flint glass in the middle and one of crown glass on each side, the edge of the crown glass prisms pointing toward the base of the flint glass prism, *i. e.*, the edges of the crown and flint glass prisms point in opposite directions. The flint glass prism serves to give the dispersion or separation into colors, while the crown glass prisms serve to make the emergent rays parallel with the incident rays, so that one looks directly into the prism along the axis of the microscope.

The Amici prism is in a special tube which is hinged to the ocular and held in position by a spring. It may be swung free of the ocular. In connection with the ocular is the slit mechanism and a prism for reflecting horizontal rays vertically for the purpose of obtaining a comparison spectrum (§ 137). Finally near the top is a lateral tube with mirror for the purpose of projecting an Angström scale of wave lengths upon the spectrum (§ 138).

§ 129. Various Kinds of Spectra.—By a spectrum is meant the colored bands appearing when light traverses a dispersing prism or a diffraction grating, or is affected in any way to separate the different wave lengths of light into groups. When daylight or some good artificial light is thus dispersed one gets the appearance so familiar in the rain-bow.

(*A*) *Continuous Spectrum.*—In case a good artificial light or the electric light is used the various rain-bow or spectral colors merge gradually into one another in passing from end to end of the spectrum. There are no breaks or gaps.

(*B*) *Line Spectrum.*—If a gas is made incandescent the spectrum it produces consists, not of the various rain-bow colors, but of sharp, narrow, bright lines, the color depending on the substance. All the rest of the spectrum is dark. These line spectra are very strikingly shown by various metals heated till they are in the form of incandescent vapor.

(*C*) *Absorption Spectrum.*—By this is meant a spectrum in which there are dark lines or bands in the spectrum. The most striking and interesting of the absorption spectra is the *Solar Spectrum*, or spectrum of sun-light. If this is examined carefully it will be found to be crossed by dark lines, the appearance be-

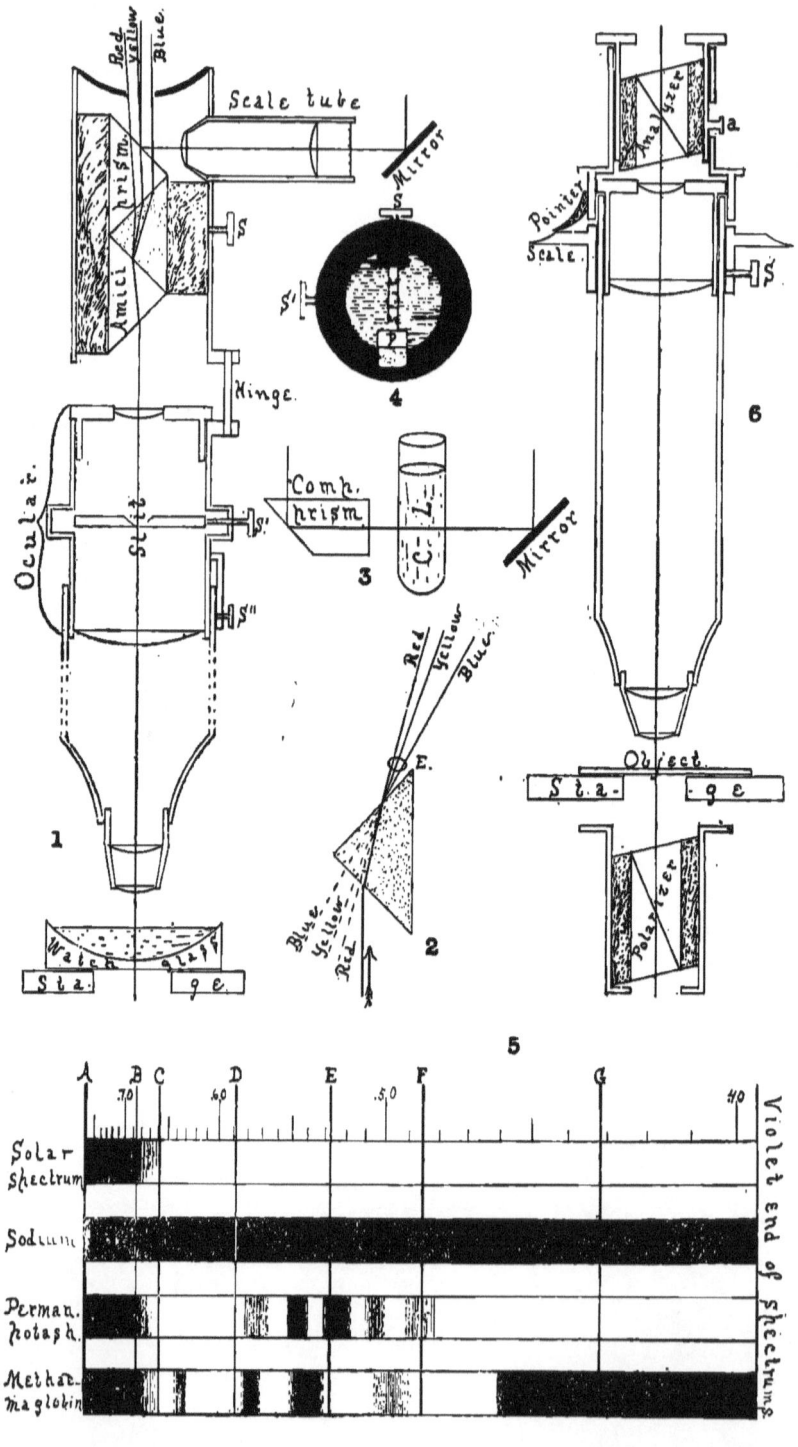

EXPLANATION OF PLATE VI.

THE MICRO-SPECTROSCOPE AND MICRO-POLARISCOPE.

Fig. 1. Section of the tube and stage of the microscope with the spectral ocular or micro-spectroscope in position.

Amici Prism (§ 128).—The direct vision prism of Amici in which the central shaded prism of flint glass gives the dispersion or separation into colors while the end prisms of crown glass cause the rays to emerge approximately parallel with the axis of the microscope (see § 128). A single ray is represented as entering the prism and this is divided into three groups (Red, Yellow, Blue), which emerge from the prism, the red being least and the blue most bent toward the base of the flint prism (see Fig. 2).

Hinge.—The hinge on which the prism tube turns when it is swung off the ocular.

Ocular (§ 128).—The ocular in which the slit mechanism takes the place of the diaphragm (§ 134). The eye-lens is movable as in a micrometer ocular, so that the slit may be accurately focused for the different colors (§ 136).

S. Screw for setting the scale of wave lengths (§ 138).

S'. Screw for regulating the width of the slit (§ 138).

S''. Screw for clamping the micro-spectroscope to the tube of the microscope.

Scale Tube.—The tube near the upper end containing the Angström scale and the lenses for projecting the image upon the upper face of the Amici prism, whence it is reflected upward to the eye with the different colored rays. At the right is a special mirror for lighting the scale. For arranging and focusing the scale, see § 138.

Slit.—The linear opening between the knife edges. Through the slit the light passes to the prism. It must be arranged parallel with the refracting edge of the prism and of such a width that the Fraunhofer or Fixed Lines are very clearly and sharply defined when the eye-lens is properly focused (§§ 134, 135, 136).

Stage.—The stage of the microscope. This supports a watch-glass with sloping sides for containing the colored liquid to be examined.

Fig. 2. *Flint-Glass Prism* showing the separation or dispersion of white light into the three groups of colored rays (Red, Yellow, Blue), the blue rays being bent the most from the refracting edge. This figure shows also that if the eye is placed at E, close to the prism, the different colored rays will appear in the direction which they reach the eye and consequently are crossed in being projected into the field of vision and the real position is inverted. The same is true in looking into the micro-spectroscope. The actual position of the different colors may be determined by placing some ground glass or some of the lens-paper (see Additions and Corrections) near the prism and observing with the eye at the distance of distinct vision.

Fig. 3. *Comparison Prism* with tube for colored liquid (C. L.), and mirror. The prism reflects horizontal rays vertically, so that when the prism is made to cover part of the slit two parallel spectra may be seen, one from light sent directly through the entire microscope and one from the light reflected upward from the comparison prism.

Fig. 4. *View of the Slit Mechanism from Below.*—Slit, the linear space between the knife edges through which the light passes.

P. Comparison prism beneath the slit and covering part of it at will.

S, S'. Screws for regulating the width and length of the slit.

Fig. 5. *Various Spectrums.*—All except that of Sodium were obtained by diffused day-light and the slit of such a width that gave the most distinct Fraunhofer lines.

It frequently occurs that with a substance giving several absorptions bands (*e. g.*, chlorophyll) the density or thickness of the solution must be varied to show all the different bands clearly.

Solar-Spectrum.—With diffused day-light and a narrow slit the spectrum is not visible much beyond the fixed line B. In order to extend the visible spectrum in the red to the line A, one should use direct sunlight and a piece of ruby glass in place of the watch glass in Fig. 1.

Sodium Spectrum.—The line spectrum (§ 129 B) of sodium obtained by lighting the microscope with an alcohol flame in which some salt of sodium is glowing. With the micro-spectroscope the sodium line seen in the solar spectrum and with the incandescent sodium appears single, except under very favorable circumstances (§ 138). By using a comparison spectrum of day-light with the sodium spectrum the light and dark D-line will be seen to be continuous as here shown.

Permanganate of Potash.—This spectrum is characterized by the presence of five absorption bands in the middle of the spectrum and is best shown by using a $\frac{1}{10}$ per cent. solution of permanganate in water in a watch glass as in Fig. 1 (§ 145).

Met-haemoglobin.—The absorption spectrum of met-haemoglobin is characterized by a considerable darkening of the blue end of the spectrum and of four absorptions bands, one in the red near the line C and two between D and E nearly in the place of the two bands of oxy-haemoglobin; finally there is a somewhat faint, wide band near F. Such a met-haemoglobin spectrum is best obtained by making a solution of blood in water of such a concentration that the two oxy-haemoglobin bands run together (§ 146), and then adding three or four drops of a $\frac{1}{10}$ per cent. aqueous solution of permanganate of potash. Soon the bright red will change to a brownish color, when it may be examined.

Fig. 6. Sectional View of a Microscope with the Polariscope in Position (§§ 150-157).

Analyzer and Polarizer.—They are represented with corresponding faces parallel so that the polarized beam could traverse freely the analyzer. If either nicol were rotated 90° they would be crossed and no light would traverse the analyzer unless some polarizing substance were used as object (§ 151). (a) Slot in the analyzer tube so that the analyzer may be raised or lowered to adjust it for difference of level of the eye-point in different oculars (§ 153).

Pointer and Scale.—The pointer attached to the analyzer and the scale or divided circle clamped (by the screw S) to the tube of the microscope. The pointer and scale enable one to determine the exact amount of rotation of the analyzer (§ 152).

Object.—The object whose character is to be investigated by polarized light.

It not infrequently occurs that colored liquids of unknown character are presented for identification or examination with the micro-spectroscope. For example, some red liquid, to determine whether or not it is blood. If the absorption bands differ markedly from the bands of haemoglobin or its derivitives, one may conclude with considerable certainty that the liquid is not colored by blood, provided always, of course, that the examination is made with sufficient care. If the bands correspond with some absorption spectrum of blood, then some reagent must be used whose action on blood solutions have been determined, to see if the reaction of the unknown liquid corresponds. (See MacMunn).

ing as if one were to draw pen marks across a continuous spectrum at various levels, sometimes apparently between the colors and sometimes in the midst of a color. These dark lines are the so-called Fraunhofer Lines. Some of the principal ones have been lettered with Roman capitals, A. B. C. D. E. F. G. H., commencing at the red end. The meaning of these lines was for a long time enigmatical, but it is now known that they correspond with the bright lines of a line spectrum (B). For example, if sodium is put in the flame of a spirit lamp it will vaporize and become luminous. If this light is examined there will be seen one or two bright yellow bands corresponding in position with D of the solar spectrum (Pl. V, Fig. 45). If now the spirit lamp-flame, colored by the incandescent sodium, is placed in the path of the electric light, and the light examined as before, there will be a continuous spectrum, except dark lines in place of the bright sodium lines. That is, the comparatively cool yellow light of the spirit lamp cuts off or absorbs the intensely hot yellow light of the electric light; and although the spirit flame sends yellow light to the spectroscope it is so faint in comparison with the electric light that the sodium lines appear dark. It is believed that in the sun's atmosphere there are incandescent metal vapors (sodium, iron, etc.) but that they are so cool in comparison with the rays of their wave length in the sun that the cooler light of the incandescent metallic vapors absorbs the light of corresponding wave length, and are, like the spirit lamp-flame, unable to make up the loss, and therefore the presence of the dark lines.

Absorption spectra from colored substances.—While the solar spectrum is an absorption spectrum, the term is more commonly applied to the spectra obtained with light which had passed through or has been reflected from colored objects which are not self-luminous.

It is the special purpose of the micro-spectroscope to investigate the spectra of colored objects which are not self luminous, as blood and other liquids, various minerals, as malazeit, etc. The spectra obtained by examining the light reflected from these colored bodies or transmitted through them, possess, like the solar spectrum, dark lines or bands, but the bands are usually much wider and less sharply defined. Their number and position depend on the substance or its constitution (Pl. V, Fig. 45), and their width, in part, upon the thickness of the body. With some colored bodies, no definite bands are present. The spectrum is simply restricted at one or both ends and various of the other colors are considerably lessened in intensity. This is true of many colored fruits.

§ 130. Angström and Stokes Law of Absorption Spectra.—The wave lengths of light absorbed by a body when light is transmitted through some of its substances are precisely the waves radiated from it when it becomes self-luminous. For example, a piece of glass that is yellow when cool, gives out blue light when it is hot enough to be self luminous. Sodium vapor absorbs two bands of yellow light (D lines); but when light is not sent through it, but itself is luminous and examined as a source of light its spectrum gives bright sodium lines, all the rest of the spectrum being dark.

§ 131. Law of Color.—The light reaching the eye from a colored, solid, liquid or gaseous body lighted with white light, will be that due to white light less the light waves that have been absorbed by the colored body. Or in other words, it will be due to the wave lengths of light that finally reach the eye from the object. For example, a thin layer of blood under the microscope will appear yellowish green, but a thick layer will appear pure red. If now these two layers are examined with a micro-spectroscope, the thin layer will show all the colors, but the red

end will be slightly, and the blue end considerably restricted, and some of the colors will appear of considerably lessened intensity. Finally there may appear two shadow-like bands, or if the layer is thick enough, two well-defined dark bands in the green (§ 146).

If the thick layer is examined in the same way, the spectrum will show only red with a little orange light, all the rest being absorbed. Thus the spectroscope shows which colors remain, in part or wholly, and it is the mixture of this remaining or unabsorbed light that gives color to the object.

§ 132. **Complementary Spectra.**—While it is believed that Angström's law (§ 130) is correct, there are many bodies on which it cannot be tested, as they change in chemical or molecular constitution before reaching a sufficiently high temperature to become luminous. There are compounds, however, like those of didymium, erbium and terbium, which do not change with the heat necessary to render them luminous, and with them the incandescence and absorption spectra are mutually complementary, the one presenting bright lines where the other presents dark ones (Daniell).

ADJUSTING THE MICROSPECTROSCOPE.

§ 133. The micro-spectroscope or spectroscopic ocular is put in the place of the ordinary ocular of the microscope, and clamped to the top of the tube by means of a screw for the purpose.

§ 134. **Adjustment of the Slit.**—In place of the ordinary diaphragm with circular opening, the spectral ocular has a diaphragm composed of two movable knife edges by which a slit-like opening of greater or less width and length may be obtained at will by the use of screws for the purpose. To adjust the slit depress the lever holding the prism-tube in position over the ocular, and swing the prism aside. One can then look into the ocular. The lateral screw should be used and the knife edges approached till they appear about half a millimeter apart. If now the Amici prism is put back in place and the microscope well lighted, one will see a spectrum by looking into the upper end of the spectroscope. If the slit is too wide, the colors will overlap in the middle of the spectrum and be pure only at the red and blue ends; and the Fraunhofer or other bands in the spectrum will be faint or invisible. Dust on the edges of the slit gives the appearance of longitudinal streaks on the spectrum.

§ 135. **Mutual arrangement of Slit and Prism.**—In order that the spectrum may appear as if made up of colored bands going directly across the long axis of the spectrum, the slit must be parallel with the refracting edge of the prism. If the slit and prism are not thus mutually arranged, the colored bands will appear oblique and the whole spectrum may be greatly narrowed. If the colored bands are oblique, grasp the prism tube and slowly rotate it to the right or to the left until the various colored bands extend directly across the spectrum.

§ 136. **Focusing the Slit.**—In order that the lines or bands in the spectrum shall be sharply defined, the eye-lens of the ocular should be

accurately focused on the slit. The eye-lens is movable, and when the prism is swung aside it is very easy to focus the slit as one focused for the ocular micrometer (§ 114). If one now uses daylight there will be seen in the spectrum the dark Fraunhofer lines (Pl. V, Fig. 45, E. F., etc.).

To show the necessity of focusing the slit, move the eye-lens down or up as far as possible, and the Fraunhofer lines cannot be seen. While looking into the spectroscope move the ocular lens up or down and when it is focused the Fraunhofer lines will reappear. As the different colors of the spectrum have different wave lengths, it is necessary to focus the slit for each color if the sharpest possible pictures are desired.

§ 137. **Comparison or Double Spectrum.**—In order to compare the spectra of two different substances it is desirable to be able to examine their spectra side by side. This is provided for in the the better forms of micro-spectroscopes by a prism just below the slit, so placed that light entering it from a mirror at the side of the drum shall be totally reflected in a vertical direction, and thus parallel with the rays from the microscope. The two spectra will be side by side with a narrow dark line separating them. If now the slit is well focused and daylight be sent through the microscope and into the side to the reflecting or comparison prism, the colored bands and the Fraunhofer dark lines will appear directly continuous across the two spectra. The prism for the comparison spectrum is movable and may be entirely thrown out of the field if desired. When it is to be used, it is thrown about half way across the field so that the two spectrums shall have about the same width.

§ 138. **Scale of Wave Lengths.**—In the Abbe micro-spectroscope the scale is in a separate tube near the top of the prism and at right angles to the prism-tube. A special mirror serves to light the scale, which is projected upon the spectrum by a lens in the scale-tube. This scale is of the Angström form, and the wave lengths of any part of the spectrum may be read off directly, after the scale is once set in the proper position, that is, when it is set so that any given wave length on the scale is opposite the part of the spectrum known by previous investigation to have that particular wave length. The point most often selected for setting the scale is opposite the sodium lines where the wave length is, according to Angström, $0.5892\,\mu$. In adjusting the scale, one may focus very sharply the dark sodium line of the solar spectrum and set the scale so that the number 0.589 is opposite the sodium or D line, or a method that is frequently used and serves to illustrate § 129 C., is to sprinkle some salt of sodium (carbonate of sodium is good) in an alcohol lamp flame and to examine this flame. If this is done in a dark-

ened place with a spectroscope, two narrow bright yellow bands will be seen in the yellow part of the spectrum. If now ordinary daylight is sent through the comparison prism, the bright lines of the sodium will be seen to be directly continuous with the dark lines at D in the solar spectrum (Pl. V, Fig. 45). Now, by reflecting light into the scale-tube, the image of the scale will appear on the spectrum, and by a screw just under the scale-tube, but in the prism-tube, the proper point on the scale (0.589 μ) can be brought opposite the sodium bands. All the scale will then give the wave lengths directly. Sometimes the scale is oblique to the spectrum. This may be remedied by turning the prism-tube slightly one way or the other. It may be due to the wrong position of the scale itself. If so, grasp the milled ring at the distal end of the scale-tube and, while looking into the spectroscope, rotate the tube until the lines of the scale are parallel with the Fraunhofer lines. It is necessary in adjusting the scale to be sure that the larger number 0.70 is at the red end of the spectrum.

The numbers on the scale should be very clearly defined. If they do not so appear, the scale-tube must be focused by grasping the outer tube of the scale-tube and moving it toward or from the prism-tube until the scale is distinct. In focusing the scale, grasp the outer scale-tube with one hand and the prism-tube with the other, and push or pull in opposite directions. In this way one will be less liable to injure the spectroscope.

§ 139. **Designation of Wave Length.**—Wave lengths of light are designated by the Greek letter λ, followed by the number indicating the wave length in some fraction of a meter. With the Abbe microspectroscope the micron is taken as the unit as with other microscopical measurements (§ 108). Various units are in use, as the one hundred thousandth of a millimeter, millionths or ten millionths of a millimeter. If these smaller units are taken, the wave lengths will be indicated either as a decimal fraction of a millimeter or as whole numbers. Thus, according to Angström, the wave length of sodium light is 5892 ten millionths mm. or 589.2 millionths, or 58.92 one hundred thousandths, or 0.5892 of one thousandth mm., or 0.5892 μ. The last would be indicated thus, $\lambda D = 0.5892\,\mu$.

§ 140. **Lighting for the Micro-spectroscope.**—For opaque objects a strong light should be thrown on them either with a concave mirror or a condensing lens. For transparent objects the amount of the substance and the depth of color must be considered. As a general rule it is well to use plenty of light, as that from an Abbe illuminator with a large opening in the diaphragm, or with the diaphragm entirely removed. For very small objects and thin layers of liquids it may be better to use less light. One must try both methods in a given case, and learn by experience. For many objects some good artificial light is better than daylight.

The direct and the comparison spectrums should be about equally illuminated. One can manage this by putting the object requiring the greater amount of illumination on the stage of the microscope, and lighting it with the Abbe illuminator.

§ 141. **Objectives to Use with the Micro-spectroscope.**—If the material is of considerable bulk a low objective (18 to 50 mm.) is to be preferred. This depends on the nature of the object under examination, however. In case of individual crystals one should use sufficient magnification to make the real image of the crystal entirely fill the width of the slit. The length of the slit may then be regulated by the screw on the side of the drum, and also by the comparison prism. If the object does not fill the whole slit the white light entering the spectroscope with the light from the object might obscure the absorption bands.

In using high objectives with the micro-spectroscope one must very carefully regulate the light (§§ 39, 68), and sometimes shade the object.

§ 142. **Focusing the Objective.**—For focusing the objective the prism-tube is swung aside, and then the slit made wide by turning the adjusting screw at the side. When the slit is open, one can see objects when the microscope is focused as with an ordinary ocular. After an object is focused, it may be put exactly in position to fill the slit of the spectroscope, then the knife edges are brought together till the slit is of the right width; if the slit is then too long it may be shortened by using one of the mechanism screws on the side, or if that is not sufficient, by bringing the comparison prism farther over the field. If one now replaces the Amici prism and looks into the microscope, the spectrum is liable to have longitudinal shimmering lines. To get rid of these focus up or down a little so that the microscope will be slightly out of focus.

§ 143. **Amount of Material Necessary for Absorption Spectra and its Proper Manipulation.**—The amount of material necessary to give an absorption spectrum varies greatly with different substances, and can be determined only by trial. If a transparent solid is under investigation it is well to have it in the form of a wedge, then successive thicknesses can be brought under the microscope. If a liquid substance is being examined, a watch glass with sloping sides forms an excellent vessel to contain it, then successive thicknesses of the liquid can be brought into the field as with the wedge shaped-solid. Frequently only a very weak solution is obtainable, in this case it can be placed in a homœopathic vial or in some glass tubing sealed at the end, then one can look lengthwise through the liquid and get the effect of a more concentrated solution. For minute bodies like crystals or blood corpuscles, one may proceed as described in the previous section.

MICRO-SPECTROSCOPE—EXPERIMENTS.

§ 144. Put the micro-spectroscope in position, arrange the slit and the Amici prism so that the spectrum shall show the various spectral colors going directly across it (§§ 133–135) and carefully focus the slit. This may be done either by swinging the prism-tube aside and proceeding as for the ocular micrometer (§ 114) or by moving the eye-lens of the ocular up and down while looking into the micro-spectroscope until the dark lines of the solar spectrum are distinct. If they cannot be made distinct by focusing the slit, then the light is too feeble or the slit is too wide (§ 136). With the lever move the comparison prism across half the field so that the two spectra shall be of about equal width.

§ 145. **Absorption Spectrum of Permanganate of Potash.**—Make a solution of permanganate of potash in water of such a strength that a stratum 3 or 4 mm. thick is transparent. Put this solution in a watch-glass with sloping sides, and put it under the microscope. Use a 50 mm. or 18 mm. objective and use the full opening of the illuminator. Light strongly. Look into the spectroscope and slowly move the watch-glass into the field. Note carefully the appearance with the thin stratum of liquid at the edge and then as it gradually thickens on moving the watch-glass still farther along. Count the absorption bands and note particularly the red and blue ends. Compare carefully with the comparison spectrum.

§ 146. **Absorption Spectrum of Blood.**—Obtain blood from a recently killed animal, or flame a needle, and after it is cool prick the finger two or three times in a small area, then wind a handkerchief or a rubber tube around the base of the finger and squeeze the finger with the other hand. Some blood will ooze out of the pricks. Rinse this off in a watch-glass partly filled with water. Continue to add the blood until the water is quite red. Place the watch-glass of diluted blood under the microscope in place of the permanganate, using the same objective, etc. Note carefully the spectrum. It would be advantageous to determine the wave length opposite the center of the dark bands. This may be done easily by setting the scale properly as described in § 138. Make another preparation, but use a homœopathic vial instead of a watch-glass. Cork the vial and lay it down upon the stage of the microscope. Observe the spectrum. It will be like that in the watch-glass. Remove the cork and look through the whole length of the vial. The bands will be very much darker and if the solution is thick enough only red and a little orange will appear. Reinsert the cork and incline the vial so that the light traverses a very thin layer then gradually elevate the vial and the effect of a thicker

and thicker layer may be seen. Note especially that the two characteristic bands unite and form one wide band as the stratum of liquid thickens. Compare with the following:

Add to the vial of diluted blood a drop or two of ammonium sulphide, such as is used for a reducing agent in chemical laboratories. Shake the bottle gently and then allow it to stand for ten or fifteen minutes. Examine it and the two bands will have been replaced by a single less clearly defined band in about the same position. The blood will also appear somewhat purple. Shake the vial vigorously and the color will change to the bright red of fresh blood. Examine it again with the spectroscope and the two bands will be visible. After five or ten minutes another examination will show but a single band. Incline the bottle so that a very thin stratum may be examined. Note that the stratum of liquid must be considerably thicker to show the absorption band than was necessary to show the two bands in the first experiment. Furthermore, while the single band may be made quite black on thickening the stratum, in will not separate into two bands with a thinner stratum. In this experiment it is very instructive to have a second vial of fresh, diluted blood, say that from the watch-glass, before the opening to the comparison prism. The two-banded spectrum will then be in position to be compared with the spectrum of the blood treated with the ammonium sulphide.

The two banded spectrum is of *oxy-haemoglobin* or arterial blood, the single banded spectrum is of *haemoglobin* (sometimes called reduced haemoglobin) or venous blood, that is the respiratory oxygen is present in the two banded spectrum but absent from the single banded spectrum. When the bottle was shaken the haemoglobin took up oxygen from the air and became oxy-haemoglobin, as occurs in the lungs, but soon the ammonium sulphide took away the respiratory oxygen thus reducing the oxy-haemoglobin to haemoglobin. This may be repeated many times. For further spectroscopic study of blood, see Part II.

§ 147. **Colored Bodies not giving Distinctly Banded Absorption Spectra.**—Some quite brilliantly colored objects, like the skin of a red apple, do not give a banded spectrum. Take the skin of a red apple, mount it on a slide, put on a cover-glass and add a drop of water at the edge of the cover. Put the preparation under the microscope and observe the spectrum. Although no bands will appear, in some cases at least, yet the ends of the spectrum will be restricted and various regions of the spectrum will not be so bright as the comparison spectrum. Here the red color arises from the mixture of the unabsorbed wave lengths, as occurs with other colored objects. In this case, however, not all the light of a given wave length is absorbed,

consequently there are no clearly defined dark bands, the light is simply less brilliant in certain regions and the red rays so preponderate that they give the prevailing color.

§ 148. **Absorption Spectrum of Colored Minerals.**—As example take some malazeit sand on a slide and either mount it in balsam (see § 176), or cover and add a drop of water. The examination may be made also with the dry sand, but it is less satisfactory. Light well with transmitted light, and move the preparation slowly around. Absorption bands will appear occasionally. Swing the prism-tube off the ocular, open the slit and focus the sand. Get the image of one or more grains directly in the slit, then narrow and shorten the slit so that no light can reach the spectroscope that has not traversed the grain of sand. The spectrum will be very satisfactory under such conditions. It is frequently of great service in determining the character of unknown mineral sands to compare their spectra with known minerals. If the absorption bands are identical, it is strong evidence in favor of the identity of the minerals.

REFERENCES TO THE MICRO-SPECTROSCOPE AND SPECTRUM ANALYSIS.

§ 149. The micro-spectroscope is playing an ever increasingly important role in the spectrum analysis of animal and vegetable pigments, and of colored mineral and chemical substances, therefore a somewhat extended reference to literature will be given. Full titles of the books and periodicals will be found in the Bibliography at the end.

Angström, Recherches sur le spectre solaire, etc. Also various papers in periodicals. See Royal Soc's Cat'l Scientific Papers; Anthony & Brackett ; Beale, p. 269; Behrens, p. 139; Behrens, Kossel und Schiefferdecker, p. 63 ; Carpenter p. 104 ; Browning, How to Work with the Spectroscope, and in Monthly Micr. Jour., II, p. 65 ; Daniell, Principles of Physics. The general principles of spectrum analysis are especially well stated in this work, pp. 435-455 ; Dippel, p. 277 ; Frey ; Gamgee, p. 91 ; Halliburton ; Hogg, p. 122 ; also in Monthly Micr. Jour., Vol. II, on colors of flowers ; Jour. Roy. Micr. Soc., 1880, 1883 and in various other vols. ; Kraus ; Lockyer ; M'Kendrick ; Macmunn ; and also in Philos. Trans. R. S., 1886 ; Various vols. of Jour. Physiol.; Nägeli und Schwendener ; Proctor ; Ref. Hand-Book Med. Sciences, Vol. I, p. 577, VI, p. 516, VII, p. 426 ; Roscoe ; Schellen ; Sorby, in Beale, p. 269, also Proc. R. S., 1874, p. 31, 1867, p. 433 ; see also in the Scientific Review, Vol. V, p. 66, Vol. II, p. 419. The larger works on Physiology, Chemistry and Physics may also be consulted with profit.

MICRO-POLARISCOPE.

§ 150. The micro-polariscope or polarizer is a polariscope used in connection with a microscope.

The most common and typical form consists of two Nicol prisms, that is two somewhat elongated rhombs of Iceland spar cut obliquely and cemented together with Canada balsam. These Nicol prisms are then mounted in such a way that

the light passes through them lengthwise, and in passing is divided into two rays of plane polarized light. The one of these rays obeying most nearly the ordinary law of refraction is called the *ordinary ray*, the one departing farthest from the law is called the *extra-ordinary ray*. These two rays are not only polarized, but polarized in planes almost exactly at right angles to each other. The Nicol prism totally reflects the ordinary ray at the cemented surface so that only the extraordinary ray is transmitted.

§ 151. Polarizer and Analyzer.—The polarizer is one of the Nicol prisms. It is placed beneath the object and in this way the object is illuminated with polarized light. The analyzer is the other Nicol and is placed at some level above the object, very conveniently above the ocular.

When the corresponding faces of the polarizer and analyzer are parallel, *i. e.*, when the faces through which the oblique section passed are parallel, light passes freely through the analyzer to the eye. If these corresponding faces are at right angles, that is if the Nicols are crossed, then the light is entirely cut off and the two transparent prisms become opaque to ordinary light. There are then, in the complete revolution of the analyzer, two points, at 0° and at 180°, where the corresponding faces are parallel and where light freely traverses the analyzer. There are also two crossing points of the Nicols, at 90° and 270°, where the light is extinguished. In the intermediate points there is a sort of twilight.

§ 152. Putting the Polarizer and Analyzer in Position.—Swing the diaphragm carrier of the Abbe illuminator out from under the illuminator, remove the disk diaphragm or open widely the iris diaphragm and place the analyzer in the diaphragm carrier, then swing it back under the illuminator. Remove the ocular, put the graduated ring on the top of the tube and then replace the ocular and put the analyzer over the ocular and ring. Arrange the graduated ring so that the indicator shall stand at 0° when the field is lightest. This may be done by turning the tube down so that the objective is near the illuminator, then shading the stage so that none but polarized light shall enter the microscope. Rotate the analyzer until the lightest possible point is found, then rotate the graduated ring till the index stands at 0°. The ring may then be clamped to the tube by the side screw for the purpose.

§ 153. Adjustment of the Analyzer.—The analyzer should be capable of moving up and down in its mounting, so that it can be adjusted to the eye-point of the ocular with which it is used. If on looking into the analyzer with parallel Nicols the edge of the field is not sharp, or if it is colored, the analyzer is not in a proper position with reference to the eye-point and should be raised or lowered till the edge of the field is perfectly sharp and as free from color as the ocular with the analyzer removed.

§ 154. Objectives to Use with the Polariscope.—Objectives of the lowest power may be used and also all intermediate forms up to a 2 mm. homogeneous immersion. Still higher Objectives may be used if desired. In general, however, the lower powers are somewhat more satisfactory. A good rule to follow in this case is the general rule in all microscopic work, "*use the power that most clearly and satisfactorily shows the object under investigation.*"

§ 155. Lighting for Micro-Polariscopic Work.—Follow the general directions given in Chapter I. It is especially necessary to shade the object so that no unpolarized light can enter the objective, otherwise the field cannot be sufficiently darkened. No diaphragm is used over the polarizer for most examinations. Direct sunlight may be used to advantage with some objects, and as a rule the object

would best be very transparent. That is, tissues, fibers, etc., should be mounted in balsam (Suffolk).

§ 156. **Purpose of a Micro-Polariscope.**—The object of a micro-polariscope is to determine, in microscopic masses, one or more of the following points: (A) Whether the body is singly refractive, mono-refringent or *isotropic*, that is optically homogeneous, as is glass and crystals belonging to the cubical system ; (B) Whether the object is doubly refractive, birefringent or *anisotropic*, uniaxial or biaxial ; (C) *Pleochromism ;* (D) The rotation of the plane of polarization, as with solutions of sugar, etc. ; (E) To aid in petrology and mineralogy ; (F) To aid in the determination of very minute quantities of crystallizable substances ; (G) For the production of colors.

For petrological and mineralogical investigations the stage of the microscope should possess a graduated rotating stage so that the object can be rotated and the exact angle of rotation determined. It is also found of advantage in investigating objects with polarized light where colors appear, to combine a polariscope and spectroscope (Spectro-Polariscope).

MICRO-POLARISCOPE—EXPERIMENTS.

§ 157. Arrange the polarizer and analyzer as directed above (§ 152) and use an 18 mm. objective except when otherwise directed.

(A) **Isotropic or Singly Refractive Objects.**—Light the microscope well and cross the Nicols, shade the stage and make the field as dark as possible (§ 151). As an isotropic substance, put an ordinary glass slide under the microscope. The field will remain dark. As an example of a crystal belonging to the cubical system and hence isotropic, make a strong solution of common salt (sodium chloride Na Cl.), put a drop on a slide and allow it to crystallize, put it under the microscope, remove the analyzer, focus the crystals and then replace the analyzer and cross the Nicols. The field and the crystals will remain dark.

(B) **Anisotropic or Doubly Refracting Objects.**—Make a fresh preparation of carbonate of lime crystals like that described for pedesis (§ 95), or use a preparation in which the crystals have dried to the slide, use a 5 or 3 mm. objective, shade the object well, remove the analyzer and focus the crystals, then replace the analyzer. Cross the Nicols. In the dark field will be seen multitudes of shining crystals, and if the preparation is a fresh one in water, part of the smaller crystals will alternately flash and disappear. By observing carefully, some of the larger crystals will be found to remain dark with crossed Nicols, others will shine continuously. This shows that the crystals are uniaxial. If the crystals are in such a position that the light passes through them parallel with the principal axis, the crystals are isotropic like the salt crystal and remain dark. If, however, the light traverses them in any other direction the ray from the polarizer is divided into two constituents vibrating in planes at right angles to each other, and one of these

will traverse the analyzer, hence such crystals will appear as if self luminous in a dark field. The experiment with these crystals from the frog succeeds well with a 2 mm. homogeneous immersion.

As further illustrations of anisotropic objects, mount some cotton fibers in balsam (Ch. V), also some of the Japanese paper (§ 72). These furnish excellent examples of vegetable fibers.

Striated muscular fibers are also very well adapted for polarizing objects.

As examples of biaxial crystals, crystallize some borax, or carbonate of lead on a slide as directed for the common salt, and use the crystals as object. As all these objects restore the light with crossed Nicols, they are sometimes called depolarizing.

(C) *Pleochromism.*—This is the exhibition of different tints as the analyzer is rotated. An excellent subject for this will be found in blood crystals (See Part II.).

(D) For the aid given by the polariscope in micro-chemistry, see Ch. V.

(E) See works on petrology and mineralogy for the application of the micro-polarizer in those subjects.

(F) For the production of gorgeous colors a plate of selenite giving blue and yellow colors is placed between the polarizer and the object. If properly mounted, the selenite is very conveniently placed on the diaphragm carrier of the Abbe illuminator, just above the polarizer.

It is very instructive and interesting to examine organic and inorganic substances with a micro-polarizer. If the objects enumerated in § 96 were all examined with polarized light an additional means of detecting them would be found.

REFERENCES TO THE POLARISCOPE AND TO THE USE OF POLARIZED LIGHT.

Anthony & Brackett ; Behrens, 133 ; Behrens, Kossel und Schiefferdecker ; Carnoy, 61 ; Carpenter, 131, 839 ; Daniell, 494 ; v. Ebener ; Gamgee ; Halliburton, 36, 272 ; Hogg, 133, 729 ; Lehmann ; M'Kendrick ; Nägeli und Schwendener, 299 ; Quekett ; Suffolk, 125 ; Valentin.

CHAPTER V.

SLIDES AND COVER-GLASSES, MOUNTING, LABELING AND STORING MICROSCOPICAL PREPARATIONS—EXPERIMENTS IN MICRO-CHEMISTRY.

APPARATUS AND MATERIAL FOR THIS CHAPTER.

Microscope, compound and simple (Ch. I); Micro-Spectroscope and polariscope (Ch. IV); Slides and cover-glasses (§§ 158, 160); Cleaning mixtures for glass (§ 164); Alcohol and distilled or filtered water (§ 161); Fine forceps for handling cover-glasses (§§ 161, 172); Old handkerchiefs or Japanese paper (§§ 72, 161); Paper boxes for storing cover-glasses (§§ 161, 163); Cover-glass measurer; Mounting material,—Farrant's solution, glycerin, glycerin-jelly and Canada balsam (§§ 188–191); Centering card and lined card for serial sections (§ 172); Net-micrometer for arranging minute objects like diatoms (§ 195); Labels, (§ 194); Carbon ink for writing labels (§ 182); Faber's pencils for writing on glass, china, etc. (§ 186); Writing diamond (§ 182); Shellac cement (§§ 169, 193); Cabinet (§ 183); Reagents for experiments in micro-chemistry (§ 196).

The laboratory furnishes all of the above articles except a simple microscope, slides and cover-glasses, fine forceps, handkerchiefs, paper boxes and Faber pencils.

SLIDES AND COVER-GLASSES.

§ 158. **Slides, Glass Slides or Slips, Microscopic Slides or Slips.**—These are strips of clear, flat glass upon which microscopic specimens are usually mounted for preservation and ready examination. The size that has been almost universally adopted for ordinary preparations is 25 x 76 millimeters (1 x 3 inches). For rock sections, slides 25 x 45 mm. or 32 x 32 mm. are used; for serial sections, slides 50 x 75 mm. or 37 x 87 mm. are used. For special purposes, slides of the necessary size are employed without regard to any conventional standard.

Whatever size of slide is used, it should be made of clear glass and the edges should be ground. It is altogether false economy to mount microscopic objects on slides with unground edges.

§ 159. **Cleaning Slides.**—For new slides a thorough rinsing in clean water with subsequent wiping with a soft towel, and then an old soft handkerchief, usually fits them for ordinary use. If they are not satisfactorily cleaned in this way, soak them a short time in 50% or 75% alcohol, let them drain for a few moments on a clean towel or on blotting paper and then wipe with a soft cloth. In handling the slides grasp them by their edges to avoid soiling the face of the slide. After the

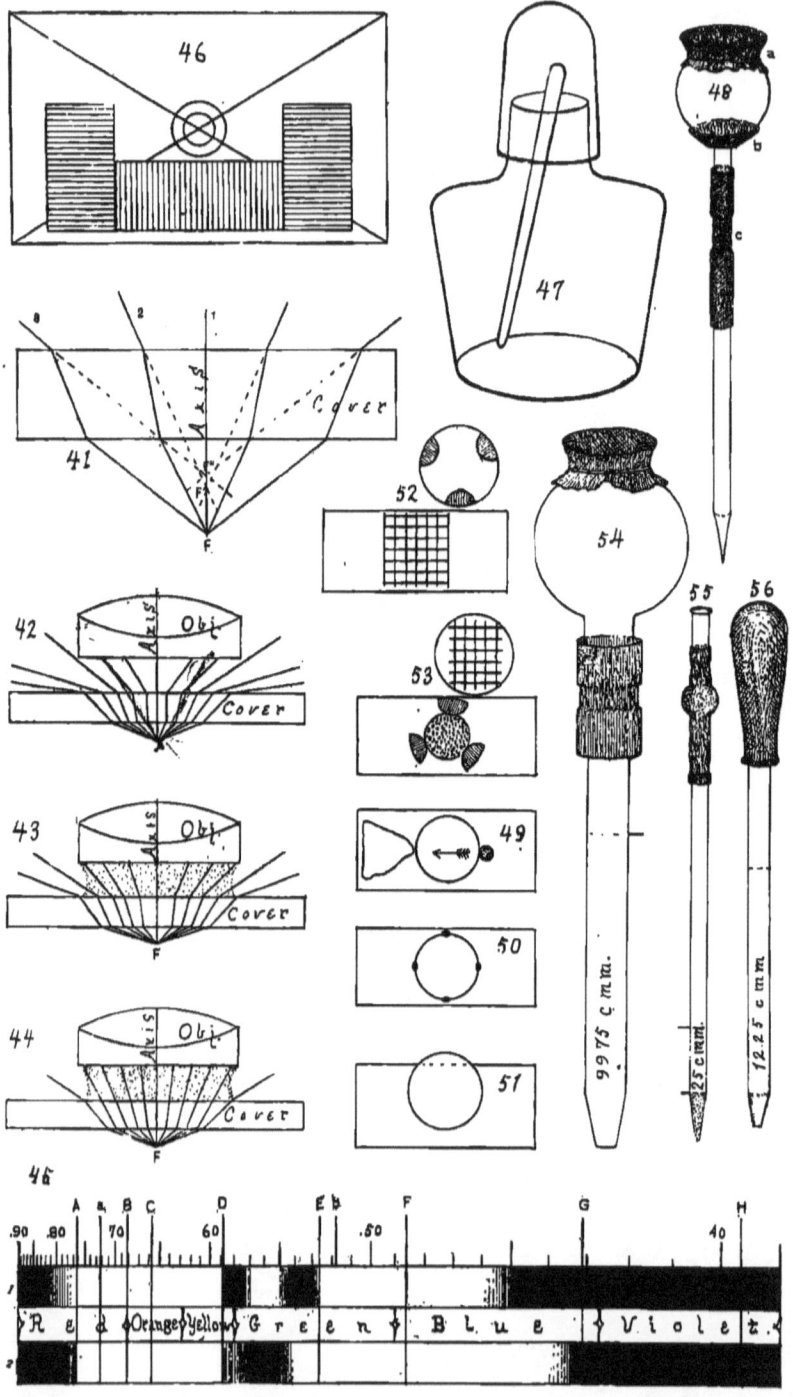

slides are cleaned, they should be stored in a place as free as possible from dust.

For used slides, if only water, glycerin or glycerin jelly has been used on them, they may be cleaned with water, or preferably, warm water and then with alcohol if necessary. Where balsam or any oily or gummy substance has been used upon the slides they may be best freed from the balsam, etc., by soaking them for a week or more, in one of the cleaning mixtures for glass (§ 164). After all foreign matter is removed the slides should be very thoroughly rinsed in water to remove all the cleaning mixture. They may then be treated as directed for new slides.

§ 160. **Cover-Glasses or Covering Glasses.**—These are circular or quadrangular pieces of thin glass used for covering and protecting microscopic objects. They should be very thin, $\frac{10}{100}$ to $\frac{25}{100}$ millimeter (see table, § 17). It is better never to use a cover glass over $\frac{20}{100}$ mm. thick then the preparation may be studied with a 2 mm. oil immersion as well as with lower objectives. Except for objects wholly unsuited for high powers, it is a great mistake to use cover-glasses thicker than the working distance of a homogeneous objective (§ 38).

The cover-glass should always be considerably larger than the object over which it is placed.

§ 161. **Cleaning Cover-Glasses.**—New cover-glasses should be put into a glass dish of some kind containing one of the cleaning mixtures (§ 164) and allowed to remain a day or longer. In putting them in, push one in at a time and be sure that it is entirely immersed, otherwise they adhere very closely, and the cleaning mixture is unable to act freely. Soiled covers should be left a week or more in the cleaning mixture. An indefinite sojourn in the cleaner does not seem to injure the slides or covers. After one day or longer, pour off the cleaning mixture into another glass jar, and rinse the cover glasses, moving them around with a gentle rotary motion. Continue the rinsing until all the cleaning mixture is removed. One may rinse them occasionally, and in the meantime allow a very gentle stream of water to flow on them or they may be allowed to stand quietly and have the water renewed from time to time. When the cleaning mixture is removed rinse the covers well with distilled water, and then cover them with 50% to 75% alcohol.

Wiping the cover-glasses.—When ready to wipe the cover-glasses remove several from the alcohol and put them on a soft dry cloth or on some of the Japanese paper, to let them drain. Grasp a cover-glass by its edges, cover the thumb and index of the other hand with a soft clean cloth or some of the Japanese paper. Grasp the cover between the thumb and index and rub the surfaces. In doing this it is necessary to keep the thumb and index well opposed or on directly opposite

faces of the cover so that no strain will come on it, otherwise the cover is liable to be broken.

When a cover is well wiped, hold it up and look through it toward some dark object. The cover will be seen partly by transmitted and partly by reflected light, and any cloudiness will be easily seen. If the cover does not look clear, breathe on the faces and wipe again. If it is not possible to get a cover clear in this way it should be put again into the cleaning mixture.

As the covers are wiped put them in a clean paper box. Handle them always by their edges, or use fine forceps. Do not put the fingers on the faces of the covers, for that will surely cloud them.

§ 162. **Cleaning Large Cover-Glasses.**—For serial sections and especially large sections, large quadrangular covers are used. These are to be put one by one into cleaning mixture as for the smaller covers and treated in every way the same. In wiping them one may proceed as for the small covers, but special care is necessary to avoid breaking them. A safe and good way to clean the large covers, is to take two perfectly flat, smooth blocks considerably larger than the cover-glasses. These blocks are covered with soft clean cloth, or with several thicknesses of the Japanese paper; if now the cover-glass is placed on the one block and rubbed with the other the cover may be cleaned as by rubbing its faces with the cloth covered finger and thumb.

§ 163. **Measuring the Thickness of Cover-Glasses.**—It is of the greatest advantage to know the exact thickness of the cover-glass on an object; for, (a) One would not try to use objectives in studying the preparation of a shorter working distance than the thickness of the cover (§ 38); (b) In using adjustable objectives with the collar graduated for different thicknesses of cover, the collar might be set at a favorable point without loss of time; (c) For unadjustable objectives the thickness of cover may be selected corresponding to that for which the objective was corrected 9 (see table § 17). Furthermore, if there is a variation from the standard, one may remedy it, in part at least, by lengthening the tube if the cover is thinner, and shortening it if the cover is thicker than the standard (§ 63).

In the so called No. 1 cover-glasses of the dealers in microsopical supplies, the writer has found covers varying from $\frac{10}{100}$ mm. to $\frac{35}{100}$ mm. To use cover-glasses of so wide a variation in thickness without knowing whether one has a thick or a thin one is simply to ignore the fundamental principles on which correct microscopic images are obtained.

It is then strongly recommended that every preparation shall be covered with a cover-glass whose thickness is known, and that this thickness should be indicated in some way on the preparation.

For the purpose of measuring cover-glasses two very convenient cover-glass measures or cover-glass testers have been devised, one made and furnished by Zeiss, Reichert, etc., in Europe, and the other, devised by Edward Bausch and furnished by the Bausch and Lomb Optical Co. of Rochester, N. Y. With either of these the cleaned covers may be very quickly and accurately measured. The different thicknesses can then be put into different boxes and properly labeled. Unless one is striving for the most accurate possible results, cover-glasses not varying more than $\frac{4}{100}$ mm. may be put in the same box. For example, if one takes $\frac{15}{100}$ mm. as a standard, covers varying $\frac{2}{100}$ mm. on each side may be put into the same box. In this case the box would contain covers of $\frac{13}{100}$, $\frac{14}{100}$, $\frac{15}{100}$, $\frac{16}{100}$, and $\frac{17}{100}$ mm.

Fig. 38. Cover-Glass Measurer (Edward Bausch).

The cover-glass is placed in the notch between the two screws, and the drum is turned by the milled head at the right till the cover is in contact with the screws. The thickness is then indicated by the knife edge on the drum, and may be read off directly in $\frac{1}{100}$th mm. or $\frac{1}{1000}$th inch. In other columns is given the proper tube-length for various unadjustable objectives ($\frac{1}{4}$, $\frac{1}{6}$, $\frac{1}{8}$, and $\frac{1}{12}$, in.) made by the Bausch and Lomb Optical Co.

§ 164. **Cleaning Mixtures for Glass.**—The cleaning mixtures used for cleaning slides and cover-glasses are those commonly used in chemical laboratories :

(A) *Dichromate of Potash and Sulphuric Acid.*
Dichromate of potash ($K_2 Cr_2 O_7$) . . . . 200 grams.
Water, distilled or ordinary . . . . 1000 cc.
Sulphuric acid ($H_2 SO_4$) . . . . 1000 cc.

Dissolve the dichromate in the water by the aid of heat. Pour the solution into a bottle that has been warmed. Add slowly and at intervals the sulphuric acid.

For making this mixture, ordinary water, commercial dichromate and strong commercial sulphuric acid should be used. It is not necessary to employ chemically pure materials.

This is a very excellent cleaning mixture and is practically odorless. It is exceedingly corrosive and must be kept in glass vessels. It may be used more than once, but when the color changes markedly from that seen in the fresh mixture it should be thrown away.

(B) *Sulphuric and Nitric Acid Mixture.*
Nitric acid ($H NO_3$) . . . . 200 cc.
Sulphuric acid ($H_2 SO_4$) . . . . 300 cc.

The acids should be strong, but they need not be chemically pure. The two acids are mixed slowly, and kept in a glass-stoppered bottle. This is a more corrosive mixture than (A) and has the undesirable feature of giving off very stifling fumes, therefore it must be carefully covered. It may be used several times. It acts more rapidly than the dichromate mixture but on account of the fumes is not so well adapted for general laboratories.

MOUNTING, AND PERMANENT PREPARATION OF MICROSCOPICAL OBJECTS.

§ 165. **Mounting a microscopical object** is so arranging it upon some suitable support (glass slide) and in some suitable mounting medium that it may be satisfactorily studied with the microscope.

Some objects are mounted dry or in air, others in some liquid miscible with water, as glycerin, and still others in some resinous medium like Canada balsam. Special methods of procedure are necessary in order to mount objects successfully in each of these ways. The best mounting medium and the best method of mounting in a given case can only be determined by experiment, unless some previous observer has already supplied the information.

The cover-glass on a permanent preparation should always be considerably larger than the object; and where several objects are put under one cover-glass it is false economy to crowd them too closely together.

§ 166. **Mounting Cells.**—Many objects are of considerable thickness and require a space or cell in which to be mounted, the wall of the cell serving to support the cover-glass and to contain the mounting medium. Where objects are mounted dry, that is in air, a cell must always be used to support the cover-glass and to prevent the soft cement used in sealing the preparation from running in by capillarity and thus flooding the preparation.

Fig. 38a. Turn-Table for Sealing Cover-Glasses and Making Shallow Mounting Cells. (Queen & Co.)

§ 167. **Preparation of Mounting Cells.**—(A) *Thin Cells.* These are most conveniently made of some of the microscopical cements. Shellac is one of the best and most generally applicable (§ 193). To prepare a shellac cell, place the slide on a turn-table (Fig. 38^a) and center it, that is get the center of the slide over the center of the turn-table. Select a guide ring on the turn-table which is a little smaller than the cover-glass to be used, take the brush from the shellac, being sure there is not enough cement adhering to it to drop. Whirl the turn-table and hold the brush lightly on the slide just over the guide ring selected. An even ring of the cement should result. If it is uneven, the cement is too thick or too thin or too much was on the brush. After a ring is thus prepared remove the slide and allow the cement to dry spontaneously or heat the slide in some way. Before the slide is used for mounting, the cement should be so dry when it is cold that it does not dent with the finger nail applied to it.

A cell of considerable depth may be made with the shellac by adding successive layers as the previous one drys.

(B) *Deep cells* are sometimes made by building up cement cells, but more frequently, paper, wax, glass, hard rubber or some metal is used for the main part of the cell; Paper rings, block tin or lead rings are easily cut out with gun punches. These rings are fastened to the slide by using some cement like the shellac.

§ 168. **Sealing the Cover-Glass.**—(A) *For dry objects mounted in cells.* When an object is mounted in a cell, the slide is warmed until the cement is slightly sticky, or a very thin coat of fresh cement is put

on. The cover-glass is warmed slightly also, both to make it stick to the cell more easily, and to expel any remaining moisture from the object. When the cover is put on it is pressed down all around over the cell until a shining ring appears, showing that there is an intimate contact. In doing this use the convex part of the fine forceps or some other blunt, smooth object; it is also necessary to avoid pressing on the cover except immediately over the wall of the cell for fear of breaking the cover. When the cover is in contact with the wall of cement all around, the slide should be placed on the turn-table and carefully arranged so that the cover-glass and cell wall will be concentric with the guide rings of the turn-table. Then the turn-table is whirled and a ring of fresh cement is painted, half on the cover and half on the cell wall (Fig. 40). If the cover-glass is not in contact with the cell wall at any point and the cell is shallow, there will be great danger of the fresh cement running into the cell and injuring or spoiling the preparation.

When the cover-glass is properly sealed, the preparation is put in some safe place for the drying of the cement. It is advisable to add a fresh coat of cement occasionally.

(B) *Thick or deep cells.* These may be made of paper, sheet lead or block tin, etc. They should be slightly deeper than the object to be mounted is thick. It is sometimes advisable to have a circular opening and an oblong wall instead of using a mere ring. In any case the cell wall is cemented to the slide and the cement well dried before use. If the cell is for dry objects or for those in glycerin, a ring of fresh cement is added just before putting on the cover-glass. If glycerin jelly, a resinous substance, or Farrant's solution is to be used as the mounting medium no cement on the top is necessary.

§ 169. Sealing the Cover-Glass when no Cell is Used.—(A) *For glycerin mounted specimens.* The superfluous glycerin is wiped away as carefully as possible with a moist cloth, then four minute drops of cement are placed at the edge of the cover (Pl. V, Fig. 50), and allowed to harden for half an hour or more. These will anchor the cover-glass, then the preparation may be put on the turn-table and a ring of cement put around the edge while whirling the turn-table.

(B) *For objects in glycerin jelly, Farrant's solution or a resinous medium.* The mounting medium is first allowed to harden, then the superfluous medium is scraped away as much as possible with a knife, and then removed with a cloth moistened with water for the glycerin jelly and Farrant's solution or with alcohol, chloroform or turpentine, etc., if a resinous medium is used. Then the slide is put on a turn-table and a ring of the shellac cement added. (C) *Balsam preparations* may be sealed with shellac as soon as they are prepared, but it is better to allow them to dry for a few days. One should never use a cement for

sealing preparations in balsam or other resinous media unless the solvent of the cement is not a solvent of the balsam, etc. Otherwise the cement will soften the balsam and finally run in and mix with it, and partly or wholly ruin the preparation. Shellac is an excellent cement for sealing balsam preparations, as it never runs in, and it serves to avoid any injury to the preparation when cedar oil, etc., are used for homogeneous immersion objectives.

§ 170. **Order of Procedure in Mounting Objects Dry or in Air.**

1. A cell of some kind is prepared. It should be slightly deeper than the object is thick (§§ 166, 167).

2. The object is thoroughly dried (desiccated) either in dry air or by the aid of gentle heat.

3. If practicable the object is mounted on the cover-glass, if not it is placed in the bottom of the cell.

4. The slide is warmed till the cement forming the cell wall is somewhat sticky, or a thin coat of fresh cement is added; the cover is warmed and put on the cell and pressed down all around till a shining ring indicates its adherence (§ 168).

5. The cover-glass is sealed (§ 168).
6. The slide is labeled (§ 179).
7. The preparation is cataloged and safely stored (§§ 181-183).

MOUNTING OF OBJECTS IN MEDIA MISCIBLE WITH WATER.

§ 171. Many objects are so greatly modified by drying that they must be mounted in some medium other than air. In some cases water or water with something in solution is used. Glycerin of various strengths, glycerin jelly and Farrant's solution are also much employed (§§ 189, 190). All these media keep the object moist and therefore in a condition resembling the natural one. The object is usually and properly treated with gradually increasing strengths of glycerin or fixed by some fixing agent (See Part II) before being permanently mounted in strong glycerin or either of the other media.

In all of these different methods, unless glycerin of increasing strengths has been used to prepare the tissue, the fixing agent is washed away with water before the object is finally and permanently mounted in either of the media.

For glycerin jelly or Farrant's solution no cell is necessary unless the object has a considerable thickness.

§ 172. **Order of Procedure in Mounting Objects in Glycerin.**

1. A cell is employed if the object is of considerable thickness.
2. The suitably prepared object (§ 171) is placed on the center of a

clean slide, and if no cell is required a centering card is employed to facilitate the centering (Pl. V, Fig. 46).

3. A drop of pure glycerin is put upon the object, or if a cell is used, enough to fill the cell.

4. In putting on the cover-glass it is grasped with fine forceps and the under side breathed on to slightly moisten it so that the glycerin will adhere, then one edge of the cover is put on the cell or slide and the cover gradually lowered upon the object (Pl. II, Fig. 14). The cover is then gently pressed down. If a cell is used, a fresh coat of cement is added before mounting (§ 168 B).

5. The cover-glass is sealed (§§ 168, 169).

6. The slide is labeled (§ 179).

7. The preparation is cataloged and safely stored (§§ 181–183).

§ 173. **Order of Procedure in Mounting Objects in Farrant's Solution.**

A cell is only necessary when the object is of considerable thickness. The object may be considerably thicker than when glycerin is used without requiring a cell.

1. The centering card is used (Pl. V, Fig. 46), and a small drop of Farrant's solution put in the middle. The suitably prepared object is then put on the solution in the center and carefully arranged as desired.

2. A drop of the Farrant's solution is put on the object, or if a cell is used it is filled with the medium.

3. The cover-glass is grasped with fine forceps, the lower side breathed upon and then it is gradually lowered upon the object (Pl. II, Fig. 14), and slightly pressed down.

4. After the mounting medium has hardened round the edge of the cover-glass the superfluous medium is scraped and wiped away and the cover sealed with shellac (§§ 168, 169).

5. The slide is labeled (§ 179).

6. The preparation is cataloged and safely stored (§§ 181–183).

§ 174. **Order of Procedure in Mounting Objects in Glycerin Jelly.**

1. Unless the object is quite thick no cell is necessary with glycerin jelly.

2. A slide is gently warmed and placed on the centering card (Pl. V, Fig. 46) and a drop of warmed glycerin jelly is put on its center. The suitably prepared object is then arranged in the center of the slide.

3. A drop of the warm glycerin jelly is then put on the object, or if a cell is used it is filled with the medium.

4. The cover-glass is grasped with fine forceps, the lower side breathed on and then gradually lowered upon the object (Pl. II, Fig. 14), and gently pressed down.

5. After mounting, the preparation is left flat in some cool place till the glycerin jelly sets, then the superfluous amount is scraped and wiped away and the cover-glass sealed with shellac (§§ 168, 169).

6. The slide is labeled (§ 179).

7. The preparation is cataloged and safely stored (§§ 181–183).

MOUNTING OBJECTS IN RESINOUS MEDIA.

§ 175. While the media miscible with water offer many advantages for mounting animal and vegetable tissues the preparations so made are liable to deteriorate. In many cases, also, they do not produce sufficient transparency to enable one to use sufficiently high powers for the demonstration of minute details.

By using sufficient care almost any tissue may be mounted in a resinous medium and retain all its details of structure.

For the successful mounting of an object in a resinous medium it must in some way be deprived of all water and all liquids not miscible with the resinous mounting medium. There are two methods of bringing this about:

(A) *By drying or desiccation.* This answers well for many objects, for example, a fly's wing, crystals, etc.

(B) *By a series of displacements.* The first step in the series is *Dehydration*, that is the water is displaced by some liquid which is miscible both with the water and the next liquid to be used. Strong alcohol (95% or stronger) is usually employed for this. Plenty of it must be used to displace the last trace of water. The tissue may be soaked in a dish of the alcohol, or alcohol from a pipette may be poured upon it. Dehydration usually occurs in the thin objects to be mounted in balsam in 5 to 15 minutes. If a dish of alcohol is used it must not be used too many times, as it loses its strength.

The second step is clearing. That is some liquid which is miscible with the alcohol and also with the resinous medium is used. This liquid is highly refractive in most cases and consequently this step is called *clearing* and the liquid a *Clearer.* The clearer displaces the alcohol, and renders the object more or less translucent. In case the water was not all removed a cloudiness will appear in parts or over the whole of the preparation. In this case the preparation must be returned to alcohol to complete the dehydration.

One can tell when a specimen is properly cleared by holding it over some dark object. If it is cleared it can be seen only with difficulty, as

but little light is reflected from it. If it is held toward the window, however, it will appear translucent.

The third and final step is the displacement of the clearer by the resinous mounting medium.

The specimen is drained of clearer and allowed to stand for a short time till there appears the first sign of dullness from evaporation of the clearer from the surface. Then a drop of the resinous medium is put on the object and finally a cover-glass is placed over it, or a drop of the mounting medium is spread on the cover and it is then put on the object.

It is in many ways more convenient to perform the series of displacements on the slide. This must be done with serial sections. If the preparations are not fastened to the slide, some workers perform the dehydration and clearing in separate dishes.

§ 176. **Order of Procedure in Mounting Objects in Resinous Media by Desiccation.**

1. The object suitable for the purpose (fly's wing, etc.) is thoroughly dried in dry air or by gentle heat.

2. The object is arranged as desired in the center of a clean slide on the centering card (Pl. V., Fig. 46.)

3. A drop of the mounting medium (§ 191) is put directly upon the object or spread on a cover-glass.

4. The cover-glass is put on the specimen with fine forceps (Pl. II, Fig. 14), but in no case does one breathe on the cover as when media miscible water are used.

5. The cover-glass is pressed down gently.

6. The slide is labeled (§ 179).

7. The preparation is cataloged and safely stored (§§ 181–183).

8. Although it is not absolutely necessary, it is better to seal the cover with shellac after the medium has hardened round the edge of the cover (§ 169 C).

§ 177. **Order of Procedure in Mounting Objects in Resinous Media by successive Displacements.**

1. A suitable object is selected, for example a section of animal tissue, and is centered on a clean slide.

2. The slide is held in the hand and the object is dehydrated by dropping upon it strong alcohol (§ 175 B).

3. The alcohol is drained from the specimen and removed by blotting paper held at the edge of the object.

4. Two or three drops of the clearer (§ 192) are put on the object to displace the alcohol (§ 175 B).

5. When the object appears translucent the clearer is drained off and blotted from the edge of the specimen (§ 175).

6. A drop of the resinous medium is put directly on the object or spread upon a cover-glass.

7. The cover-glass is put upon the object and pressed down. It may then be heated gently.

8. The slide is labeled (§ 179).

9. The preparation is cataloged and safely stored (§§ 181-183).

10. After the resin has hardened round the edge of the cover the superfluous material may be cleaned away and the cover-glass sealed with shellac. This is not absolutely necessary, but is desirable (§ 169 C).

LABELING, CATALOGING AND STORING MICROSCOPICAL PREPARATIONS.

§ 178 Every person possessing a microscopical preparation is interested in its proper management; but it is especially to the teacher and the investigator that the labeling, cataloging and storing of microscopical preparations are of importance. "To the investigator, his specimens are the most precious of his possessions, for they contain the facts which he tries to interpret, and they remain the same while his knowledge, and hence his power of interpretation, increase. They thus form the basis of further or more correct knowledge; but in order to be safe-guides for the student, teacher, or investigator, it seems to the writer that every preparation should possess two things; viz., a label and a catalog or history. This catalog should indicate all that is known of a specimen at the time of its preparation, and all of the processes by which it is treated. It is only by the possession of such a complete knowledge of the entire history of a preparation that one is able to judge with certainty of the comparative excellence of methods, and thus to discard or improve those which are defective. The teacher, as well as the investigator, should have this information in an accessible form, so that not only he but his students can obtain at any time all necessary information concerning the preparations which serve him as illustrations and them as examples."

§ 179. Labeling Ordinary Microscopical Preparations.—The label (§ 191) should possess at least the following information :—

EXAMPLE.

(1) The number of the preparation.	No. 475.
(2) The thickness of the cover-glass.	Cover-Glass, .15 mm.
(3) The name and source of the preparation.	Striated Muscular Fibers, Sartorius of Cat.
(4) The date on which the preparation is made.	October 1, 1891.

SERIAL SECTIONS.

For serial sections with collodion imbedded objects it is a great advantage to have the imbedding mass unsymmetrically trimmed, so that if a section is accidentally turned over it may be easily noticed and rectified.

Furthermore it is imperatively necessary that the object be so imbedded that the cardinal aspects, - dextral and sinistral, dorsal and ventral, cephalic and caudal, shall be known with certainty.

UPPER EDGE OF SLIDE.

LEFT END.						
	1	2	3	4	5	Series No. 75. Cover .15 mm. Slide No. 1. Transections of a Diemyctylus Embryo. Sections 1-10. Total thickness of Sections, 1 mm.
	6	7	8	9	10	
						May 20, 1892.

Fig. 57—Labeled Slide of Serial Sections.

¿ 180 b. Labeling Serial Sections : The label of a slide on which serial sections are mounted should contain at least the following : (1) The number of the series ; (2) The number of the slide in the series (if the series required more than one slide) ; (3) Kind of sections (transections, etc.) and the name of the object from which derived ; (4) The number of the first and last section on the slide ; (5) The total thickness of all the sections on the slide ; (6) The date of the series.

The combined thickness of the sections on a slide is easily determined by noting carefully the position of the microtome screw at the first and last sections, and measuring the elevation. Then if the sections are uniform the thickness of each may be easily found. The average thickness may be easily determined in any case.

¿ 181. Cataloging Preparations.—It is believed from personal experience, and from the experience of others, that each preparation (each slide or each series) should be accompanied by a catalog containing at least the information suggested in the following formula : This formula is very flexible, so that the order may be changed, and numbers not applicable in a given case may be omitted. With many objects, especially embryos and small animals, the time of fixing and hardening may be months or even years earlier than the time of imbedding. So, too, an object may be sectioned a long time after it was imbedded, and finally the sections may not be mounted at the time they are cut. It would be well in such cases to give the date of fixing under 2, and under 5, 6 and 8, the dates at which the operations were performed if they differ from the original date and from one another. In brief, the more that is known about a preparation the greater its value.

General Formula for Cataloging Microscopical Preparations:

1. The general name and source.
2. The number and date of the preparation and the name of the preparator.
3. The special name of the preparation and the common and scientific name of the object from which it is derived.
4. The age and condition of the object from which the preparation is derived.
5. The chemical treatment, — the method of fixing, hardening, dissociating, etc.
6. The mechanical treatment, — imbedded, sectioned, dissected with needles, etc.
7. The staining agent and the time required for staining.
8. Dehydrating and clearing agent, mounting medium, cement used for sealing.
9. The objectives and other accessories (micro-spectroscope, polarizer, etc.) for studying the preparation.
10. Remarks, including references to original papers, or to good figures and descriptions in books.

A Catalog Card Written According to this Formula:

1. Striated Muscular Fibers. Cat.
2. No. 475, (Drr. IX) Oct. 1, 1891. S. H. G., Preparator.
3. Tendinous and intra-muscular terminations of striated muscular fibers from the *Sartorius* of the cat (*Felis domestica*).
4. Cat eight months old, healthy and well nourished.
5. Muscle pinned on cork with vaselined pins and placed in 20 per cent. nitric acid immediately after death by chloroform. Left 36 hours; temperature 20° C. In alum water (sat. aq. sol.) 1 day.
6. Fibers separated on the slide with needles.
7. Stained 5 minutes with Delafield's haematoxylin.
8. Mounted in glycerin jelly (§ 174).
9. Use 18 mm. for the general appearance of the fibers, then 2 or 3 mm. objective for the details of structure (§ 75). Try the micro-polariscope (§ 157).
10. The nuclei or muscle corpuscles are very large and numerous; many of the intra-muscular ends are branched. See S. P. Gage, Proc. Amer. Micr. Soc., 1890, p. 132; Ref. Hand-Bk. Med. Sci., Vol. V., p. 59.

§ 182. General Remarks on Catalogs and Labels.—It is especially desirable that labels and catalogs shall be written with some imperishable ink. Some form of water-proof carbon ink is the most available and satisfactory. The water-proof India ink, or the engrossing carbon ink of Higgins, answers very well. As purchased the last is too thick for ordinary writing and should be diluted with one third its volume of water and a few drops of strong ammonia added.

If one has a writing diamond it is a good plan to write a label with it on one end of the slide. It is best to have the paper label also, as it can be more easily read.

The author has found stiff cards, postal card size like those used for cataloging books in public libraries, the most desirable form of catalog. A specimen that is for any cause discarded has its catalog card destroyed. New cards may then be added in alphabetical order as the preparations are made. In fact a catalog on cards has all the flexibility and advantages of the slip system of notes (see Wilder & Gage, p. 45).

80   MOUNTING AND LABELING.

Some workers prefer a book catalog. Very excellent book catalogs have been devised by Alling and by Ward (Jour. Roy. Micr. Soc., 1887, pp. 173, 348 ; Amer. Monthly Micr. Jour., 1890, p. 91 ; Amer. Micr. Soc. Proc., 1887, p. 233).

The fourth section in the cataloging formula has been introduced, as there is coming to be a belief that the tissues of young and of old animals differ in respect to the size of the nuclei (see Minot in Proc. Amer. Assoc. Adv. of Science, 1890, pp. 271-289). It is also extremely desirable to know whether the animal is well or ill nourished, healthy or diseased.

CABINET FOR MICROSCOPICAL PREPARATIONS.

§ 183. While it is desirable that microscopical preparations should be properly labeled and cataloged, it is equally important that they should be protected from injury. During the last few years several forms of cabinets or slide holders have been devised. Some are very cheap and convenient where one has but a few slides. For a laboratory or for a private collection where the slides are numerous the following characters seem to the writer essential :

(1). The cabinet should allow the slides to lie flat, and exclude dust and light.

(2). Each slide or pair of slides should be in a separate compartment. At each end of the compartment should be a groove or bevel, so that upon depressing either end of the slide the other may be easily grasped (Fig. 40). It is also desirable to have the floor of the compartment grooved so that the slide rests only on two edges, thus preventing soiling the slide opposite the object.

(3). Each compartment or each space sufficient to contain one slide of the standard size should be numbered, preferably at each end. If the compartments are made of sufficient width to receive two slides, then the double slides so frequently used in mounting serial sections may be put into the cabinet in any place desired.

(4). The drawers of the cabinet should be entirely independent, so that any drawer may be partly or wholly removed without disturbing any of the others.

(5). On the front of each drawer should be the number of the drawer in Roman numerals, and the number of the first and last compartment in the drawer in Arabic numerals (Fig. 39).

FIG. 39.

Fig. 39.—Cabinet for Microscopical Specimens, showing the method of arrangement and of numbering the drawers and indicating the number of the first and last compartment in each drawer. It is better to have the slides on which the drawers rest somewhat shorter, then the drawer-front may be entire and not notched as here shown.

MOUNTING AND LABELING.

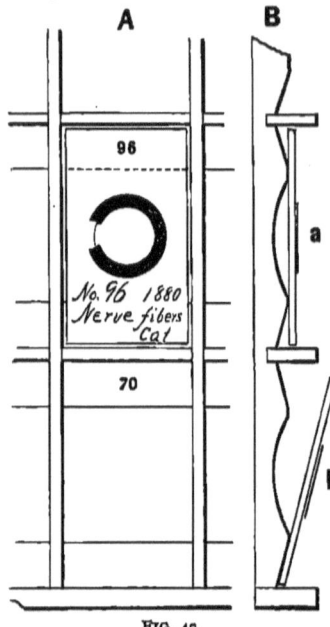

Fig. 40, A.—Part of a cabinet drawer seen from above. In compartment No. 96 is represented a slide lying flat. The label of the slide and the number of the compartment are so placed that the number of the compartment may be seen through the slide. The sealing cement is removed at one place to show that in sealing the cover-glass, the cement is put partly on the cover and partly on the slide (§ 168).

Fig. 40, B.—This represents a section of the same part of the drawer. (a) Slide resting as in A No. 96. The preparation is seen to be above a groove in the floor of the compartment. (b) One end of the slide is seen to be uplifted by depressing the other into the bevel.

FIG. 40.

MOUNTING OBJECTS—EXPERIMENTS.

§ 184. Mounting Dry, or in Air (§ 170).—Prepare a shallow cell and dry it (§ 167). Select a clean cover-glass slightly larger than the cell. Pour upon the cover a drop of a 10 per cent. solution of salicylic acid in 95 per cent. alcohol. Let it dry spontaneously. Warm the slide till the cement ring or cell is somewhat sticky, then warm the cover gently and put it on the cell, pressing down all around (§ 170). Seal the cover, label and catalog (§§ 179, 181).

A preparation of mammalian red blood corpuscles may be made very satisfactorily by spreading a very thin layer of fresh blood on a cover with the end of a slide. After it is dry, warm gently to remove the last traces of moisture and mount precisely as for the crystals. One can get the blood as directed for the Micro-spectroscopic work (§ 146).

§ 185. Mounting in Glycerin Jelly.—For this select some stained and isolated muscular fibers. Arrange them on the middle of a slide, using the centering card, and mount in glycerin jelly as directed in § 174. Air bubbles are not easily removed from glycerin jelly preparations, so care should be taken to avoid them.

§ 186. Mounting in Balsam by Desiccation (§ 176).—Find a fresh fly, or if in winter procure a dead one from a window sill or a spider's web. Carefully remove the fly's wings, being especially careful to keep them the dorsal side up. With a camel's hair brush remove any dirt that may be clinging to them. Place a clean slide on the centering card, then with fine forceps put the two wings within one of the guide rings. Leave one dorsal side up, turn the other ventral side up. Spread some Canada Balsam on the face of the cover-glass and with the fine forceps place the cover upon the wings (Pl. II, Fig. 14). Probably some air-bubles will appear in the preparation, but if the slide is put in a warm place these will soon disappear. Label, catalog, etc., (§ 176, 179, 181).*

§ 187. Mounting in Balsam by Displacement (§§ 175, 177).—For this experiment select a stained section of any organ or tissue, as the skin, or myel (spinal cord), then proceed exactly as described in §§ 175, 177.

---

*The Faber's pencils for writing on glass, china, etc., are very convenient for writing temporary labels, etc., on slides and bottles.

PREPARATION OF MOUNTING MEDIA.

§ 188. **Glycerin.**—One should procure pure glycerin for a mounting medium. It needs no preparation, except in some cases it should be filtered through filter paper or absorbent cotton to remove dust, etc.

For preparing objects for final mounting, glycerin 50 cc., water 50 cc., form a good mixture. For many purposes the final mounting in glycerin is made in an acid medium, viz., Glycerin 99 cc., Glacial acetic or formic acid, 1 cc.

By extreme care in mounting and by occasionally adding a fresh coat to the sealing of the cover-glass, glycerin preparations last a long time. They are liable to be very disappointing, however. In mounting in glycerin care should be taken to avoid air-bubbles, as they are difficult to get rid of. A specimen need not be discarded unless the air-bubbles are large and numerous.

§ 189. **Glycerin Jelly.**—Soak 25 grams of the best dry gelatin in cold water in a small agate-ware dish. Allow the water to remain until the gelatin is softened. It usually takes about half an hour. When the gelatin is softened, as may be readily determined by taking a little in the fingers, pour off the superfluous water and drain well to get rid of all the water that has not been imbibed by the gelatin. Warm the softened gelatin over a water bath and it will melt in the water it has absorbed. Add to the melted gelatin about 5 cc. of egg albumen (white of egg); stir it in well and then heat the gelatin in the water bath for about half an hour. Do not heat above 75° or 80° C., for if the gelatin is heated too hot it will be transformed into meta-gelatin and will not set when cold. The heat will coagulate the albumen and form a kind of floculent precipitate which seems to gather all fine particles of dust, etc., leaving the gelatin perfectly clear. After the gelatin is clarified it should be filtered through a hot filter and mixed with an equal volume of glycerin and 5 grams of chloral hydrate and shaken thoroughly. If it is allowed to remain in a warm place (*i. e.*, in a place where the gelatin remains melted) the air-bubbles will rise and dissapear.

In case the glycerin jelly remains fluid or semi-fluid at the ordinary temperature (18°-20° C.), the gelatin has either been transformed into meta-gelatin by too high temperature or it contains too much water. The amount of water may be lessened by heating at a moderate temperature over a water bath in an open vessel. This is a very excellent mounting medium. Air-bubbles should be avoided in mounting as they do not disappear.

§ 190. **Farrant's Solution.**—Take 25 grams of clean, dry, gum arabic; 25 cc. of a saturated aqueous solution of arsenious acid; 25 cc. of glycerin. The gum arabic is soaked for several days in the arsenic water, then the glycerin is added and carefully mixed with the dissolved or softened gum arabic.

This medium retains air-bubbles with great tenacity. It is much easier to avoid than to get rid of them in mounting. For the method of mounting in this see § 173.

§ 191. **Canada Balsam, Balsam of Fir.**—This is one of the oldest and most satisfactory of the resinous media used for mounting microscopical preparations. Sometimes it is used in the natural state, but experience has shown that it is better to get rid of the natural volatile constituents. A considerable quantity, half a liter or more, of the natural balsam is poured into shallow plates in layers about 1 or 2 centimeters thick, then the plates are put in a warm, dry place, on the back of a stove or on a steam radiator, and allowed to remain until the balsam may be powdered when it is cold. This requires a long time, the time depending on the temperature and the thickness of the layer of balsam.

MOUNTING AND LABELING.

When the volatile products have evaporated, the balsam is broken into small pieces or powdered in a mortar and mixed with about an equal volume of xylol, turpentine or chloroform. It will dissolve in this and then should be filtered through absorbent cotton or a filter paper, using a paper funnel.* The balsam is too thin in this condition for mounting, but so made for the sake of filtering it. After it is filtered it is evaporated slowly in an open dish or a wide-mouth bottle or jar till it is of a syrupy consistency at the ordinary temperature. It is then poured into a bottle with a glass cap like a spirit lamp. For use it is put into a small spirit lamp (Pl. V, Fig 47).

The xylol is much the best substance to use for thinning the balsam. Such *xylol balsam*, as it is then called, may be used for mounting any object suitable for balsam mounting. The dehydration must be very perfect, however, as xylol is wholly immiscible with water.

§ 192. **Clearing Mixture** (§ 175).—One of the most satisfactory and generally applicable clearers is made by mixing carbolic acid crystals (*Acidum carbolicum, A. phenicum crystallizatum*) 40 cc. with rectified oil of turpentine (*Oleum terebinthinae rectificatum*) 60 cc.

§ 193. **Shellac Cement.**—Shellac cement for sealing preparations and for making shallow cells (§§ 167, 168) is prepared by adding scale or bleached shellac to 95 per cent. alcohol. The bottle should be filled about half full of the solid shellac then enough 95 per cent. alcohol added to fill the bottle nearly full. The bottle is shaken occasionally and then allowed to stand until a clear stratum of liquid appears on the top. This clear, supernatant solution is then filtered through absorbent cotton, using a paper funnel (§ 189), into an open dish or a wide-mouth bottle. To every 50 cc. of this filtered shellac 5 cc. of castor oil and 5 cc. of Venetian turpentine are added to render the shellac less brittle. This filtered shellac will be too thin and must be allowed to evaporate till it is of the consistency of thin syrup. It is then put into a capped bottle and for use into a small spirit lamp (Pl. V, Fig. 47). In case the cement gets too thick add a small amount of 95 per cent. alcohol or some thin shellac.

§ 194. **Liquid Gelatin.**—Gelatin or clear glue 75 to 100 grams. Commercial acetic acid (No. 8) 100 cc., Water 100 cc., 95 per cent. alcohol 100 cc. Glycerin 15 to 30 cc. Crush the glue and put it into a bottle with the acid, and set in a warm place, and shake occasionally. After three or more days add the other ingredients. This solution is excellent for fastening paper to glass, wood or paper. The brush must be mounted in a quill or wooden handle. For labels, it is best to use linen paper of moderate thickness. This should be coated with the liquid gelatin and allowed to dry. The labels may be cut of any desired size and attached by simply moistening them as in using postage stamps.

Very excellent blank labels are now furnished by dealers in microscopical supplies, so that it is unnecessary to prepare them one's self except for special purposes.

---

*For filtering balsam and all resinous and gummy materials, the writer has found a paper funnel the most satisfactory. It can be used once and then thrown away. Such a funnel may be very easily made by rolling a sheet of thick writing paper in the form of a cone and cementing the paper where it overlaps, or winding a string several times around the lower part. Such a funnel is best used in one of the rings for holding funnels.

ARRANGING AND MOUNTING MINUTE OBJECTS.

§ 194. Minute objects like diatoms and the scales of insects may be arranged in geometrical figures or in some fanciful way either for ornament or more satisfactory study. To do this the cover-glass is placed over the guide. This guide for geometrical figures may be a net-micrometer or a series of concentric circles. In order that the objects may remain in place, however, they must be fastened to the cover-glass. As an adhesive substance, liquid gelatin (§ 194) thinned with an equal volume of 50 per cent. acetic acid answers well. A very thin coating of this is spread on the cover with a needle or in some other way and allowed to dry. The objects are then placed on the gelatinized side of the cover and carefully got into position with a mechanical finger, made by fastening a cat's whisker in a needle holder. For most of these objects a simple microscope with stand (Fig. 8) will be found of great advantage. After the objects are arranged, one breathes very gently on the cover-glass to soften the gelatin. It is then allowed to dry and if a suitable amount of gelatin has been used, and it has been properly moistened the objects will be found firmly anchored. In mounting, one may use Canada Balsam or mount dry on a cell (§§ 170, 176). See Newcomer, Amer. Micr. Soc's Proc. 1886, p. 128; see also E. H. Griffith and H. L. Smith, Amer. Journal of Micros., iv, 102, V, 87; Amer. Monthly Micr. Jour., i, 66, 107, 113. Cunningham, The Microscope, viii, 1888, p. 237.

MICRO-CHEMISTRY AND CRYSTALLOGRAPHY—EXPERIMENTS.

§ 196. The student of science and especially chemistry so frequently requires a knowledge of the appearance of minute crystals to aid in the determination of an unknown substance or for his information in studying objects where crystals are liable to occur, that a few experiments have been introduced to give him a start in preparing and permanently mounting some of the common crystals.

It is recommended that the crystals be made in several ways, that is from alcoholic solutions, aqueous solutions saturated and dilute, by spontaneous drying and crystallization and by rapid crystallization by the aid of heat. The modifications in crystallization under these different methods of treatment are frequently very striking.

In every case the student is advised to study the appearance of the crystals in the "mother liquor." As a rule their characteristics are more clearly shown in the "mother liquor" than under any other conditions.

It is of very great advantage to examine all crystalline forms with polarized light (§ 156).

§ 197. *Determination of the Character of the Solid Sediment in Water.*—Take some of the sediment from a filter or allow a considerable volume of water to stand in a tall glass vessel to deposit its sediment. Take a concentrated drop of this sediment and mount it on a slide under a cover-glass. Study the preparation with the microscope. Probably there will be an abundance of animal and vegetable life as well as of solid sediment. Put a drop of dilute sulphuric acid (*Acidum sulphuricum dilutum, i. e.*, strong sulphuric acid 1 gram, water 9 grams) at the edge of the cover and at the opposite edge a small piece of the Japanese paper (Pl. V, Fig. 49). The acid will gradually diffuse, and if the solid particles are carbonate of lime, minute bubbles will be seen to be given off. If they are silica or clay no change will result. Sulphuric acid is recommended for this, as the mi-

croscope would be far less liable to injury than as if some acid giving off fumes were used.

§ 198. **Herapath's Method of Determining Minute Quantities of Quinine.**—For a so-called test fluid 12 c.c. of glacial acetic acid, 4 c.c. of 95 per cent. alcohol and 7 drops of dilute sulphuric acid (§ 197) are mixed. A drop of the test fluid is put on a slide and a very minute amount of quinine added. After this is dissolved, add an extremely minute drop of an alcoholic solution of iodine. "The first effect is the production of the yellow cinnamon-colored compound of iodine and quinine which forms as a small circular spot; the alcohol separates in little drops, which by a sort of repulsive movement, drive the fluid away; after a time, the acid liquid again flows over the spot, and the polarizing crystals of sulphate of iodoquinine are slowly produced in beautiful rosettes. This succeeds best without the application of heat." Dr. Herapath used this method to determine the presence of quinine in the urine of patients under quinine treatment. See Hogg, p. 150; Quarterly Jour. Micr. Sc., vol. ii, pp. 13–18. For further papers on micro-chemistry by Dr. Herapath, see the Royal Society's Catalog of Scientific Papers.

§ 199. **List of Substances for the Study of Crystallography with the Microscope.***—The substances are crystallized on the cover glass in all cases, and in all cases, except where otherwise stated, a saturated aqueous solution of the substance was first prepared.

1. Ammonium chloride; 2. Ammonium copper chloride; 3. Barium chloride; 4. Cobalt chloride (Beautiful crystals obtained by mixing the saturated aqueous solution with an equal volume of 95 per cent. alcohol. Crystallization in a current of dry air some distance above an alcohol or Bunsen flame; Mount in xylol balsam (§§ 176, 191). 5. Copper acetate; Mount dry (§ 170). 6. Copper sulphate. Crystals much more satisfactory when examined in the "mother liquor." 7. Lead nitrate; 8. Mercuric chloride (Corrosive sublimate), mount in xylol balsam (§§ 176, 191). 9. Nickel nitrate; obtain crystals by heating. Mount in xylol balsam (§§ 186, 191); 10. Potash alum; 11. Potassium chlorate; 12. Potassium dichromate. Compare specimen crystallized by heat and spontaneously; mount dry or in xylol balsam (§§ 170, 176). 13. Potassium iodide. Dilute with one or two vols. water and crystallize by heat. 14. Potassium nitrate; 15. Potassium oxalate; 16. Potassium sulphate; 17. Salicine. Fuse the dry salicine on the cover-glass, mount dry (§ 170); 17. Salicylic acid. Make a 10 per cent. solution in 95 per cent. alcohol. Let it crystallize spontaneously in the air. Mount dry (§ 170); 18. Sodium chloride (common salt). Mix sat. aq. sol. with one or two volumes of water, and heat. Mount dry or in balsam (§§ 170, 176).

§ 200. For directions and hints in micro-chemical work and crystallography, consult the various volumes of the Journal of the Roy. Micr. Soc., Zeitschrift für physiologische Chemie and other chemical journals; Wormly; Klément & Regnard; Carpenter; Hogg; Behrens Kossel und Schiefferdecker; Frey.

---

*Most of the chemicals here named were suggested to the writer by Prof. L. M. Dennis of the Chemical Department.

BIBLIOGRAPHY.

The books and periodicals named below in alphabetical order, are in the laboratory or the University library. They pertain wholly or in part to the microscope microscopical or histological methods. They are referred to in the text by initial letters or by fuller, recognizable abbreviations.

For current microscopical and histological literature, the Journal of the Royal Microscopical Society, the Index Medicus, the Zoologischer Anzeiger, and the Zeitschrift für wissenschaftliche Mikroskopie, Anatomischer Anzeiger, Biologisches Centralblatt and Physiologisches Centralblatt, taken together furnish nearly a complete record.

References to books and papers published in the past may be found in the periodicals just named, in the Index Catalog of the Surgeon General's library; in the *Royal Society's Catolog of Scientific Papers*, and in the bibliographical references given in special papers.

BOOKS.

Angström.—Recherches sur le spectre solaire, spectre normal du soleil. Upsala, 1868.

Anthony, Wm. A., and Brackett, C. F.—Elementary text-book of physics. 7th ed. Pp. 527, 165 Fig. New York, 1891.

Bausch, E.—Manipulation of the microscope. Pp. 95, illustrated. Rochester, 1891.

Beale, L. S.—How to work with the microscope. 5th ed. Pp. 518, illustrated. London, 1880. Structure and methods.

Beauregard, H., at Galippe, V.—Guide de l'élève et du praticien pour les travaux pratiques de micrographie, comprenant la technique et les applications du microscope à l'histologie végétale, à la physiologie, à la clinique, à la hygiène et à la médecine légale. Pp. 904, 570 Fig. Paris, 1880.

Behrens, J. W.—The microscope in botany. A guide for the microscopical investigation of vegetable substances. Translated and edited by Hervey and Ward. Pp. 466, illustrated. Boston, 1885.

Behrens, W., Kossel, A., und Schiefferdecker, P.—Das Mikroskop und die Methoden der mikroskopischen Untersuchung. Pp. 315, 193 Fig. Braunschweig.

Browning, J.—How to work with the micro-spectroscope. Referred to in Beale and Carpenter.

Carnoy, J. B., Le Chanoine.—La Biologie Cellulaire; Etude comparée de la cellule dans les deux règnes. Illustrated (incomplete). Paris, 1884. Structure and methods.

Carpenter, W. B.—The microscope and its revelations. 6th ed. Pp. 882, illustrated. Loudon and Philadelphia, 1881. Methods and structure.

Cooke, M. C.—One thousand objects for the microscope. Pp. 123. London, no date. 500 figures and brief descriptions of pretty objects for the microscope.

Daniell, A.—A text-book of the principles of physics. Pp. 653, 254 Fig. London, 1884.

Dippel, L.—Grundzüge der allgemeinen Mikroskopie. Pp. 524, 245 Fig. Braunschweig, 1885. Excellent discussion of the microscope and accessories.

Ebner, V. v.—Untersuchungen über die Ursachen der Anisotropie organischer Substanzen. Leipzig, 1882. Large number of references.

Ellenberger, W.—Handbuch der vergleichenden Histologie und Physiologie der Haussäugethiere. Berlin, 1884+.

Fol, H.—Lehrbuch der vergleichenden mikroskopischen Anatomie, mit Einschluss der vergleichenden Histologie und Histogenie. Illustrated (incomplete). Leipzig, 1884. Methods and structure.

Foster, Frank P.—An illustrated encyclopaedic medical dictionary, being a dictionary of the technical terms used by writers on medicine and the collateral sciences in the Latin, English, French and German languages. Illustrated, quarto volumes. Vol. I, 1888; Vol. II, 1890; Vol. III, in press.

Frey, H.—The microscope and microscopical technology. Translated and edited by G. R. Cutter. Pp. 624, illustrated. New York, 1880. Methods and structure.

Frey, H.—Hand-book of the histology and histo-chemistry of man. Translated by Barker. Illustrated. New York, 1875. Structure and chemistry. Also the 5th German edition. Leipzig, 1876.

Gamgee, A.—A text-book of the physiological chemistry of the animal body. Part I, pp. 487, 63 Fig. London and New York, 1880. Structure and methods.

Gibbs, H.—Practical histology and pathology. Pp. 107. London, 1880. Methods.

Goodale, G. L.—Physiological botany. Pp. 499+36, illustrated. New York, 1885. Structure and methods.

Halliburton, W. D.—A text-book of chemical physiology and pathology. Pp. 874, 104 illus. London and New York, 1891.

Hogg, J.—The microscope, its history, construction and application. New edition, illustrated. Pp. 764. London and New York, 1883. Much attention paid to the polariscope.

James, F. L.—Elementary microscopical technology. Part I, the technical history of a slide from the crude material to the finished mount. Pp. 107, illustrated. St. Louis, 1887.

Klément and Regnard.—Réactions microchemiques à cristaux et leur application en analyse qualitative. Pp. 126, 8 plates. Bruxelles, 1886.

Kraus, G.—Zur Kenntniss der Chlorophyllfarbstoffe. Stuttgart, 1872.

Le Conte, Joseph.—Sight—an exposition of the principles of monocular and binocular vision. Pp. 275, illustrated. New York, 1881.

Lee, A. B.—The microtomist's vade-mecum. A hand-book of the methods of microscopic anatomy.

Lehmann, C. G.—Physiological chemistry. 2 vols. Pp. 648+547, illustrated. Philadelphia, 1855.

Lehmann, O.—Molekularphysik mit besonderer Berücksichtigung mikroskopischer Untersuchungen und Anleitung zu solchen, sowie einem Anhang über mikroskopische Analyse. 2 vols. Illustrated. Leipzig, 1888–1889.

Lockyer, J. N.—The spectroscope and its application. Pp. 117, illustrated. London and New York, 1873.

M'Kendrick, J. G.—A text-book of physiology. Vol. I, general physiology. Pp. 516, 318 illus. New York, 1888.

Macdonald, J. D.—A guide to the microscopical examination of drinking water. Illustrated. London, 1875. Methods and descriptions.

MacMunn, C. A.—The spectroscope in medicine. Pp. 325, illustrated. London, 1885.

Mayall, Jr , John.—Cantor lectures on the microscope, delivered before the society for the encouragement of arts, manufactures and commerce. Nov.-Dec., 1885. (History of the microscope, and figures of many of the forms used at various times).

Nägeli und Schwendener.—Das Mikroskop, Theorie und Auwendung desselben. 2d ed. Pp. 647, illustrated. Leipzig, 1877.

Phin, J.—Practical hints on the selection and use of the microscope for beginners. 6th edition. Illustrated. New York, 1890.

Preyer, W.—Die Blutkrystalle. Jena, 1871. Full bibliography to that date.

Pringle, A.—Practical photo-micrography. Pp. 193, illustrated. New York, 1890.

Proctor, R. A.—The spectroscope and its work. London, 1882.

Prudden, T. M.—A manual of practical normal histology. Pp. 265. 2d ed. New York, 1885. Methods and structure.

Queckett, J.—A practical treatise on the use of the microscope, including the different methods of preparing and examining animal, vegetable and mineral structures. Pp, 515, 12 plates. 2d ed. London, 1852.

Ranvier, L.—Traité technique d'histologie. Pp. 1109, illustrated. Paris, 1875-1888. Structure and methods. Also German translation 1888.

Reference Hand-Book of the medical sciences. Albert H. Buck, editor. 8 quarto vols. Illustrated with many plates, and figures in the text. New York, 1885-1889.

Richardson, J. G.—A hand-book of medical microscopy. Pp. 333, illustrated. Philadelphia, 1871. Methods and descriptions.

Robin, Ch.—Traité du microscope et des injections. 2d ed. Pp. 1101, illustrated, Paris, 1877. Methods and structure.

Roscoe, Sir Henry.—Lectures on spectrum analysis. 4th. ed. London, 1885.

Rutherford, W.—Outlines of practical histology. 2d ed. Illustrated. Pp. 194. London and Philadelphia, 1876. Methods and structures.

Satterthwaite, F. E. (editor).—A manual of histology. Pp. 478, illustrated. New York, 1881. Structure and methods.

Schäfer, E. A.—A course of practical histology, being an introduction to the use of the microscope. Pp. 304, 40 Fig. Philadelphia, 1877. Methods.

Schellen, H.—Spectrum analysis, translated by Jane and Caroline Lassell. Edited with notes by W. Huggins. 13 plates, including Angström's and Kirchhoff's maps. London, 1885.

Science Lectures at South Kensington. 2 vols. Pp. 290 and 344, illustrated. One lecture on microscopes and one on polarized light. London, 1878-1879.

Seiler, C.—Compendium of microscopical technology. A guide to physicians and students in the preparation of histological and pathological specimens. Pp. 130, illustrated. New York, 1881.

Silliman, Benj., Jr.—Principles of physics, or natural philosophy. 2d edition, rewritten. Pp. 710, 722 illustrations. New York and Chicago, 1860.

Stowell, Chas. H.—The students' manual of histology, for the use of students, practitioners and microscopists. 3d ed. Pp. 368, illustrated. Ann Arbor, 1884. Structure and methods.

Strasburger, E.— Das botanische Practicum. Anleitung zum Selbststudium der mikroskopischen Botanik, für Anfänger und Fortgeschrittnere. Pp. 664, illustrated. Structure and methods. Also English translation.

Suffolk, W. T.—On microscopical manipulation. 2d ed. Pp. 227, illustrated. London, 1870.

Suffolk, W. T.—Spectrum analysis applied to the microscope. Referred to in Beale.

Trelease, Wm.—Poulsen's botanical micro-chemistry, an introduction to the study of vegetable histology. Pp. 118. Boston, 1884. Methods.

Valentin, G.—Die Untersuchung der Pflanzen und der Thiergewebe in polarisirtem Licht. Leipzig, 1861.

Vierordt.—Die quantitative Spectralanalyse in ihrer Anwendung auf Physiologie, Chemie und Technologie. Tübingen, 1874.

Whitman, C. O.—Methods of research in microscopical anatomy and embryology. Pp. 255, illustrated. Boston, 1885.

Wilder and Gage.—Anatomical Technology as applied to the domestic cat. An introduction to human, veterinary and comparative anatomy. Pp. 575, 130 Fig. 2d ed. New York and Chicago, 1886.

Wood, J. G.—Common objects for the microscope. Pp. 132. London, no date. Upwards of 400 figures of pretty objects for the microscope, also brief descriptions and directions for preparation.

Wormly, T. G.—The micro-chemistry of poisons. 2d ed. Pp. 742, illustrated. Philadelphia, 1885.

Wythe, J. H.—The microscopist, a manual of microscopy and a. compendium of microscopical science. 4th ed. Pp. 434, 252 Fig. Philadelphia, 1880.

See also Watt's Chemical dictionary, and the various general and technical en cyclopaedias.

PERIODICALS:*

The American journal of microscopy and popular science. New York, 1876–1881. Illustrated. Methods and structure.

The American monthly microscopical journal. Illustrated. 1880+. Methods and structure.

American naturalist. Illustrated. Salem and Philadelphia, 1867+. Methods and structure.

American quarterly microscopical journal, containing the transactions of the New York microscopical society. Illustrated. New York, 1878. Structure and methods.

American society of microscopists. Proceedings. Illustrated. 1878+. Methods and structure.

Anatomischer Anzeiger. Centralblatt für die gesammte wissenschaftliche Anatomie. Amtliches Organ der anatomischen Gesellschaft. Herausgegeben von Dr. Karl Bardeleben. Jena, 1886+. Besides articles relating to the microscope or histology, a full record of current anatomical literature is given.

Archiv für mikroscopische Anatomie. Illustrated. Bonn, 1865+. Structure and methods.

Centralblatt für Physiologie. Unter Mitwirkung der physiologischen Gesellschaft zu Berlin. Herausgegeben von S. Exner und J. Gad. Leipzig und Wien,

---

*NOTE.—When a periodical is no longer published, the dates of the first and last volumes are given; but if still being published, the date of the first volume is followed by a plus sign.

1887+. Brief extracts of papers having a physiological bearing. Full bibliography of current literature.

Index Medicus. New York, 1879+. Bibliography, including histology and microscopy.

Journal of anatomy and physiology. Illustrated. London and Cambridge, 1867+. Structure and methods.

Journal de micrographie. Illustrated. Paris, 1877+. Methods and Structure.

Journal of the New York microscopical society. Illustrated. New York, 1885+. Methods and structure.

Journal of Physiology. Illustrated. London and Cambridge, 1878+.

Journal of the American Chemical Society. New York, 1879+.

Journal of the Chemical Society. London, 1848+.

Journal of the royal microscopical society. Illustrated. London, 1878+. Bibliography of works and papers relating to the microscope, microscopical methods and histology. It also includes a summary of many of the papers.

The Lens, a quarterly journal of microscopy and the allied natural sciences, with the transactions of the state microscopical society of Illinois. Illustrated. Chicago, 1872-1873. Methods and structure.

The Microscope. Illustrated. Trenton, N. J., 1881+. Methods and structure.

Microscopical Bulletin, and science news. Illustrated. Philadelphia, 1883+. The editor, Edward Pennock, introduced the term "par-focal" for oculars (see vol. iii, p. 31, also, the note to § 48, p. 18).

Monthly microscopical journal. Illustrated. London, 1869-1877.

Nature. Illustrated. London, 1869+.

Philosophical Transactions of the Royal Society of London. Illustrated. London, 1665+.

Proceedings of the Royal Society. London, 1854+.

Quarterly journal of microscopical science. Illustrated. London, 1853+. Structure and methods.

Science Record. Boston, 1883-4. Methods and structure.

Zeitschrift für Instrumentenkunde.

Zeitschrift für physiologische Chemie. Strassburg, 1877+.

Zeitschrift für wissenschaftliche Mikroskopie und für mikroskopische Technik. Illustrated. Braunsch. 1884+. Methods and bibliography.

Besides the above-named periodicals, articles on the microscope or the application of the microscope appear occasionally in nearly all of the scientific journals. One is likely to get references to these articles through the Jour. Roy. Micr. Soc. or the Zeit. wiss. Mikroskopie.

INDEX.

A

Abbe camera lucida, 48; arrangement of, 49; drawing with, 51; hinge for prism, 52; inclined microscope with, 50; laboratory microscope with, 28.
Abbe condenser, 19.
Abbe illuminator, 19; experiments, 20, laboratory microscope with, 28; light, axial and oblique, 20; mirror with, 20.
Aberration, chromatic, 4; spherical, 4.
Absorption spectra, 54, 55, 56; amount of material necessary and its proper manipulation, 59; Angström and Stokes' law of, 55; banded, not given by all colored objects, 61; of blood, 60; of colored minerals, 62; of permangate of potash, 60.
Achromatic objectives, 3, 4; triplet, 2.
Achromatism, 4.
Adjustable objectives, 5; and micrometry, 46.
Adjusting collar, graduation of, 23.
Adjustment of analyzer, 63; of objective, 5, 22; of objectives for coverglass, specific directions, 23; with graduated collar, 23.
Aerial image, 13.
Air bubbles, 31; with central and oblique illumination, 31.
Air and oil, distinguished optically, 32; by reflected light, 32.
Amici prism, 54.
Ampl fier, 39.
Amplification of microscope, 36.
Analyzer, 63; adjustment and putting in position, 63.
Angle of aperture, 6, 7.
Angström and Stokes' law of absorption spectra, 55.
Angular aperture, 6, 7.
Anisotropic, 64.
Aperture of objective, 6, 7; angular, 6, 7; formula for, 7; numerical, 7.
Aplanatic objectives, 4.
Apochromatic objectives, 4.
Apparatus and material, 1, 29, 36, 54, 66.
Appearances, interpretation, 29.
Arranging and mounting minute objects, 84.
Axial light, 16; experiments, 19; with Abbe illuminator, 20.
Axial point, 6; ray, 16.
Axis, optic, 6, 16.

B

Back combination or system of objective, 3, 5.
Bacillus tuberculosis, 25.
Balsam, Canada, preparation of, 82; removal from lenses, 27.
Banded absorption spectra not given by all colored objects, 61.
Birefringent, 64.
Blood, absorption spectrum of, 60; or other albuminous material, removal, 26.
Bread crumbs, examination of, 35.
Brownian movement, 34.
Brunswick black, removal from lenses, 27.
Bubble, air, 31.
Burning point, 2.
Butterfly scales, 35.

C

Cabinet for microscopical preparations, 80.
Camera lucida, Abbe, 48, arrangement of, 49, drawing with, 51, hinge for prism, 52, with inclined microscope, 50;
Camera lucida, definition, 47; Wollaston's, 38, 48.
Canada balsam, preparation of, 82; removal from lenses, 27.
Carbonate of lime, pedesis, 35.
Card, centering, 74.
Care of eyes, 27; microscope, mechanical parts, 25, optical parts, 26; water immersion objectives, 24.
Carmine to show currents and pedesis, 34.
Catalogs and labels, ink for, 79.
Cataloging, formula, 79; preparations, 79.
Cells, mounting, 71.
Centering and arrangement of illuminator, 19, 20;
Centering card, 74.
Central light, 16; with a mirror, 19.
Chromatic aberration, 4; correction, 4.
Chemical focus, 4; rays, 4.
Cleaning back lens of objective, 27; homogeneous objectives, 25; mixtures for glass, 70; slides and cover glasses, 66–7;

Clearer, clearing, 75.
Clearing mixture, preparation of, 83.
Clothes moth, examination of scales, 35.
Cloudiness, of objective and ocular, how to determine, 29, 30; removal, 26.
Coarse adjustment of microscope, 12.
Collective, 9.
Collodion for coating glass rod, 33.
Color images, 22, 25; law of, 55.
Colored minerals, absorption spectra of, 62; substances, spectra of, 55.
Coma, 23.
Combination of lenses, back and front, 3.
Comparison prism, 57; spectrum, 57.
Compensating ocular, 4, 9.
Complementary spectra, 56.
Compound microscope, see under microscope.
Concave lenses, 4; mirror, use of, 17.
Condenser, 1; Abbe, 19; optic axis of, 20.
Condensing lens, 20.
Continuous spectrum, 54.
Contoured, doubly, 33.
Converging lens, 1, 3; lens-system, 3.
Convex lenses, 4.
Corn starch, examination of, 35.
Correction, chromatic, 4.
Cotton, examination of, 35.
Cover-glass, or covering glass, 67; adjustment, specific directions, 23; adjustment and tube length, 24; cleaning, 67, 68; larger than object, 67; measurer, 69; measuring thickness of, 68; non-adjustable objectives, table of thickness, 6; No. 1, variation of thickness, 68; putting on, 74; sealing, 71–72; tester, 69; thickness of, 5; wiping, 67.
Currents in liquids, 34.
Crystals from frog for pedesis, 35.
Crystallization under microscope, 22.
Crystallography, 84; list of substances for, 84.

D

Damar, removal from lenses, 27.
Dark-ground illumination with Abbe illuminator, 21; with mirror, 21.
Dehydration, 75.
Desiccation in mounting objects in resinous media, 75, 76.
Designation of oculars, 10.
Determination of magnification, 38.
Diaphragms and their employment, 16; effect of one too small, 30; iris, 16; ocular, 12; pin-hole, 20; size and position of opening, 16;
Diffraction grating, 54.
Direct light, 15; vision spectro-scope, 54.
Dispersing prism, 54.
Displacements, in mounting objects in resinous media, 75–76.

Distance, standard at which the virtual image is measured, 39; working d. of simple microscope or objective, 15; working d. of compound microscope, 15.
Distinctness of outline, 32.
Distortion in drawing, avoidance of 48; spherical, 4.
Dividers, measuring spread of, 37.
Double spectrum, 57.
Doubly contoured, 33.
Draw-tube, pushing in, 18.
Drawing with Abbe camera lucida, 51, 52; distance at which done (250 mm. more or less), 40; distortion, avoidance of, 48; with microscope, 47; regulating size of, 52; scale and enlargement, 53; size of, and magnification of microscope with Abbe camera lucida, 52.
Dry objectives, 4; for laboratory microscope, 28.
Dust of living rooms, examination of, 35; on objectives and oculars, how to determine, 29, 30; removal, 26.

E

Erect image, 1.
Equivalent focal length, 3; focus of objectives, 3; focus of oculars, 10.
Examination of dust of living rooms, bread crumbs, corn starch, fibres of cotton, linen, silk, human and animal hairs, potato, rice, scales of butterflies and moths, wheat, 35.
Experiments, Abbe illuminator, 20; with adjustable and immersion objectives, 22–25; compound microscope, 10; homogeneous immersion objective, 24; lighting and focusing, 17; with micro-spectroscope, 60; with micro-polarizer, 64; in mounting, 81; simple microscope, 2.
Extraordinary ray of polarized light, 63.
Eyes, care of, 27; effect on magnification, 40.
Eye-lens of the ocular, 8, 9.
Eye-piece, 8; micrometer, 42.
Eye-point, 2; of ocular, demonstration, 14.
Eye-shade, Ward's, 27, double, Pl. II.

F

Farrant's solution, in mounting objects, order of procedure, 74; preparation of, 82.
Feather, examination of, 35.
Fibers, examination of, 35.
Field, 2; with orthoscopic ocular, 12; with periscopic ocular, 12; of view with microscope, 11, 36, 47.

INDEX. 93

Field-lens, 8, 9; action of, 14; of ocular, 8.
Filter paper, Japanese, 26.
Filtering balsam, etc., paper funnel for, 83.
Fluid, immersion, 4.
Focal distance or point, principal, 3; length, equivalent, 3;
Focus, 2; always up, 18; chemical, 4; of objectives, equivalent, 3; of oculars, equivalent, 10; optical, 4; principal, 1-3.
Focusing, 2, 15; with compound microscope, 15; experiments, 17; with high objectives, 18; with low objectives, 17; objective for micro-spectroscope, 59; with simple microscope, 15; slit of micro-spectroscope, 56.
Fraunhofer lines, 55.
Front combination or lens of objective, 3; system, 5.
Function of objective, 12, 13; of ocular, 13, 14.

G

Gelatin, liquid, preparation of, 83.
Glass, cleaning mixture for, 70; rod appearance under, microscope, 33; slides or slips, 66
Glue, liquid, preparation of, 83.
Glycerin, mounting objects in, order of procedure, 73; removal, 26.
Glycerin jelly, mounting objects in, order of procedure, 74; preparation of, 82.
Gold size, removal from lenses, 27.
Goniometer ocular, 9.
Graduation of adjusting collar, 23.
Grating, diffraction, 54.
Ground glass, preparation of, 12.

H

Hæmaglobin, 61; reduced h., 61.
Hairs, examination of, 35.
Herapath's method of determining minute quantities of quinine, 85.
High oculars, 5, 9.
Highly refractive, 33.
Homogeneous immersion objectives, 4; cleaning, 25; experiments, 24; for laboratory microscope, 28.
Homogeneous liquid, 25.
Huygenian ocular, 9.

I—J

Illumination for Abbe camera lucida, 51; artificial, 15, 20; central with air and oil, 31; dark-ground, 16; dark-ground, with Abbe illuminator, 21; dark-ground, with mirror, 21; oblique, with air and oil, 32; for Wollaston's camera lucida, 48.

Illuminator, 1; Abbe, 19; Abbe, axial and oblique light, 20; Abbe, experiments, 20; Abbe, mirror and light for, 20; centering and arrangement, 19, 20; immersion, 20.
Image, aerial, 13; color, 22, 25; erect, 1; inverted, 1; inverted, real of objective, 13; real, 1; real, inverted, 3; refraction, 22, 25.
Immersion fluid, 4; illuminator, 20; liquid, 4; objective, 4, 24.
Incandescence spectra, 56.
Incident light, 15.
Index of refraction of medium in front of objective, 7.
Ink for labels and catalogs, 79.
Interpretation of appearances under the microscope, 29.
Inverted image, 1.
Iris diaphragm, 16.
Isotropic, 64.
Japanese filter or tissue paper, 12, 13, 26.

L

Labels and catalogs, ink for, 79; preparation of, 79.
Labeling microscopical preparations, 77, serial sections, 78.
Laboratory compound microscope, 27.
Lamp-light, 20.
Lenses, combination of, 3; concave, 4; condensing, 20; converging, 1, 3; convex, 4; systems of, 3.
Lens-system, 1; converging, 3.
Letters in stairs, 30.
Light with Abbe illuminator, 20; axial, 16; axial with Abbe illuminator, 20; direct, 15; central, 16; incident, 15; oblique, 16; oblique, experiments, 19; oblique with Abbe illuminator, 20; polarized, 63; reflected, incident or direct, 15; transmitted, 16; wave length of, 58.
Lighting, 15; for Abbe camera lucida, 51; artificial, 15; experiments, 17; axial, experiments, 19; kind of light, 15; for micro-polariscope, 63; for micro-spectroscope, 58; with a mirror, 17;
Line spectrum, 54.
Linen, examination of, 35.
Liquid, currents in, 34; homogeneous, 25; immersion, 4.

M

Magnification, effect of adjusting objective, 46; determination of, 38; expressed in diameters, 36; method of binocular or double vision in obtaining, 36-37; of microscope, 36; of microscope with Abbe camera lucida,

52; of microscope, compound, 37; of microscope, simple, 36; relation to eyes, 40; varying with compound microscope, 39.
Magnifier, 2.
Magnifying power of microscope, 36.
Malezeit sand, spectrum of, 62.
Marking objects, 11, note.
Material and apparatus, 1, 29, 36, 54, 66.
Measurer, cover-glass, 69.
Measuring the spread of dividers, 37.
Mechanical parts of compound microscope, 3; of laboratory microscope, 28; of microscope, care of, 25.
Mechanical stage, 28.
Medium, mounting, 5, 7.
Micro-chemistry, 84.
Micrometer, filling lines of, 37; object or objective, 37; ocular or eye-piece, 10, 42, 43; ocular, micrometry with, 44; ocular, valuation of, 43; ocular, varying valuation of, 44; ocular, ways of using, 45; stage, 37.
Micrometry, definition, 40; with adjustable objectives, 46: comparison of methods, 46–47; with compound microscope, 41; by dividing the size of image by magnification of microscope, 42; limit of accuracy in, 46; with ocular micrometer, 44; with simple microscope, 41; remarks on, 46; by stage micrometer and camera lucida, 42; by stage micrometer on which is mounted the object, 42; unit of measure in, 41.
Micro-millimeter, 41.
Micron, 41; for measuring wave-length of light, 58.
Micro-polariscope, 35, 62; for laboratory microscope, 28; lighting for, 63; objectives to use with, 63; purpose of, 64.
Microscope, definition, 1; adjustment, 12; amplification of, 36; care of, 25; field of, 11; focusing, 15; magnification, 36; polarizing, 34.
Microscope compound, definition, 1; drawing with, 47; experiments with, 10; focusing with, 15; for laboratory, 27; magnification or magnifying power, 36, 37; magnification and size of drawing with Abbe camera lucida, 52; mechanical parts of, 3; micrometry with, 41; optic axis of, 16; optical parts of, 3; polarizing, pedesis with, 34; stand of, 1; varying magnification, 39.
Microscope, simple, definition, 1; experiments with, 2; focusing with, 15; magnification of, 36; micrometry with, 41.
Microscopic objective, 3; ocular, 8; slides or slips, 66.

Microscopical preparations, cabinet for, 80; cataloging, 79; labeling, 77; tube-length, 5.
Micro-spectroscope, 54; adjusting, 56; experiments, 60; for laboratory microscope, 28; lighting for, 58; objectives to use with, 59; slit-mechanism of, 54.
Micrum, 41.
Mikron, 41.
Minerals, colored, absorption spectra of, 62.
Mirror, 1; for Abbe illuminator, 20; concave, use of, 17; light with, central and oblique, 19; lighting with, 17; plane, use of, 17; position of concave, 17.
Molecular movement, 34.
Mono-refringent, 64.
Mounting cells, preparation of, 71; medium, 5, media, preparation of, 82.
Mounting objects, dry in air, order of procedure, 73; examples in air, glycerin jelly and balsam. 81; in Farrant's solution, order of procedure, 74; in glycerin, order of procedure, 73; in glycerin jelly, order of procedure, 74; in media miscible with water, 73; microscopical objects, 70; in resinous media, by drying or desiccation, order of procedure, 75, 76; in resinous media by successive displacements, order of procedure, 75, 76.
Movement, Brownian, or, molecular, 34.
Myopia, effect on magnification, 40.

N

Negative oculars, 8, 9.
Net-micrometer, 50.
Nicol prism, 62.
Nomenclature of objectives, 3.
Non-achromatic objectives, 4.
Non-adjustable objectives, 5; thickness of cover-glass for, table, 6.
Nose-piece, 11.
Numerical aperture of objectives, 7.

O

Object micrometer, 37; putting under microscope, 11; having plane or irregular outlines, relative position in a microscopical preparation, 30; transparent with curved outlines, relative position in microscopic preparations, 30; shading, 25.
Objective, 1; achromatic, 3, 4; adjustable, 5; adjustable, experiments, 22; adjustment for, 22; aerial image of, 13; aperture of, 6, 7; aplanatic, 4;

INDEX.   95

apochromatic, 4; back combination of, 3; cleaning back lens of, 27; collar, graduated for adjustment, 23; compound, 7; cloudiness or dust, how to determine, 30; dry, 4; equivalent focus of, 3; focusing for microspectroscope, 59; front combination of, 3; function of, 12, 13; high, focusing with, 18; homogeneous immersion, 4; homogeneous immersion, cleaning, 25; homogeneous immersion, experiments, 22, 24; immersion, 4; index of refraction of medium in front of, 7; inverted, real image of, 13; for laboratory microscope, 27, 28; lettering, 3; of low and medium power, 4; low, focusing with, 17; micrometer, 37; to use with micro-polariscope, 63; microscopic, 3; to use with micro-spectroscope, 59; for micro-spectroscope, focusing, 59; nomenclature of, 3; non-achromatic, 4; non-adjustable, 5; non-adjustable, thickness of coverglass for, table, 6; numbering, 3; oil-immersion, 4; putting in position and removing, 10; single-lens, 7; terminology of, 3; unadjustable, 5; water immersion, 24; water immersion, experiments, 22; working distance of, 15.
Oblique light, 16; with Abbe illuminator, 20; experiments, 19; with a mirror, 19.
Ocular, 1; achromatic, 8; aplanatic, 8; binocular, 8; cloudiness, how to determine, 29; Campani's, 8; compensating, 4, 9; compound, 8; deep, 8; designation by magnification or combined magnification and equivalent focus, 10; dust, how to determine, 29; equivalent focus of, 10; erecting, 8; eye-point of, demonstration, 14; field-lens, 8; focus, equivalent of, 10; function of, 13, 14; goniometer, 9; high, 5, 9; holosteric, 9; Huygenian, 9; Kellner's, 9; lettering of, 10; low, 9; micrometer, 9, 10, 42, 43; micrometer, micrometry with, 44; micrometer, putting in position, 43; micrometer, valuation of, 43; micrometer, varying valuation, 44; micrometer, ways of using, 45; micrometric, 9; microscopic, 8, 9; negative, 8, 9; numbering, 10; orthoscopic, 9, 12; par-focal, 18; periscopic, 9, 12; positive, 8, 9; projection, 9; projection, designation of, 10; putting in position and removing, 10; Ramsden's, 9; searching, 9; shallow, 9; solid, 9; spectral, 9, 10, 54; spectroscopic, 9, 10, 54; stereoscopic, 8; working, 9.

Oil and air, appearances and distinguishing optically, 31, 32.
Oil-globules, with central illumination, 31; with oblique illumination, 31, 32.
Oil-immersion objectives, 4.
Optic axis, 6; of condenser, or illuminator, 20; of microscope, 16.
Optical, combination, 1; focus, 4; parts, 1; parts of compound microscope, 3; parts of microscope, care of, 26; section, 33.
Order of procedure in mounting objects, dry or in air, 73; in Farrant's solution, 74; in glycerin, 73; in glycerin jelly, 74; in resinous media by desiccation, 76; in resinous media by successive displacements, 76.
Ordinary ray with polarizer, 63.
Orthoscopic ocular, field with, 12.
Outline, distinctness of, 32.
Oxy-hæmoglobin, 61.

P

Paper, bibulous, filter or Japanese, 12, 13; for cleaning oculars and objectives, 26.
Paraffin, removal from lenses, 27.
Parfocal oculars, 18.
Parts, optical and mechanical of microscope, 1.
Pedesis, 34; compared with currents, 34; with polarizing microscope, 34; proof of reality, 35.
Periscopic ocular, field with, 12.
Permanganate of potash, absorption spectrum of, 60.
Pin-hole diaphragm, 20.
Photography, 4.
Plane mirror, use of, 17.
Pleochromism, 64, 65.
Pleurasigma angulatum, 19.
Point, axial, 6.
Polarized light, 63; extraordinary and ordinary ray of, 63.
Polarizer, 62; and analyzer, putting in position, 63.
Polarizing microscope, pedesis with, 34.
Position of objects or parts of same object, 30.
Positive oculars, 8, 9.
Potato, examination of, 35.
Power of microscope, 36.
Preparation of Canada balsam, Farrant's solution, glycerin, glycerin jelly, 82.
Preparation of clearing mixture, liquid gelatin and shellac cement, 83.
Presbyopia, effect on magnification, 40.
Price of American and foreign microscope, 28.
Principal focus, 1, 2, 3; focal distance, 3; point, 2.

INDEX

Prism of Abbe camera lucida, 49–50; Amici, 54; comparison, 57; dispersing, 54; Nicol, 62; reflecting, 57; and slit of micro-spectroscope, mutual arrangement, 56.
Projection ocular, 9; designation of, 10.
Pumice stone for pedesis, 34.
Pushing in draw-tube, 18.
Putting on cover-glass, 74; an object under microscope, 11.

Q—R

Quinine, Herapath's method of determining minute quantities of, 85.
Ray, chemical, 4; ordinary of polarized light, 63; extraordinary, 63.
Real image, 1; inverted, 3.
Reflected light, 15.
Reflecting prism, 57.
Refraction images, 25; index of medium in front of objective, 7.
Refractive, doubly, 64; highly, 63; singly, 64.
Resinous media, mounting objects in, order of procedure, by drying or desiccation, 75; by series of displacements, 75, 76.
Revolver, 11.
Rice, examination of, 35.
Rule or scale for magnification and micrometry, 37.

S

Scale, of magnification and micrometry, 37; of wave lengths, 57.
Scales of butterflies and moths, examination of, 35.
Screen of ground glass, 12, 13.
Sealing cover-glass, 71, 72.
Section, optical, 33; serial, 28, 78.
Sediment in water, determination of character, 84.
Selenite plate for polariscope, 65.
Serial sections, 28; arranging and labeling, 78; determining thickness of, 78.
Shading object, 25; for micro-polariscope, 35.
Shellac cement, preparation of, 83; removal from lenses, 27.
Sight, injury or improvement in microscopic work, 27.
Silk, examination of, 35.
Simple microscope, see under microscope.
Slides, 66; cleaning, 66.
Slips, 66.
Slit mechanism of micro-spectroscope, 54; adjusting and focusing, 56; slit and prism, mutual arrangement, 56.
Solar spectrum or s. of sunlight, 54.
Spectral, colors, 4; ocular, 10, 54.
Spectroscope, direct vision, 54.
Spectroscopic ocular, 10, 54.
Spectrum, 4, 54.

Spectrum, absorption, 54; amount of material necessary and its proper manipulation, 59, Angström and Stokes' law of, 55; banded, not given by all colored objects, 61; of blood, 60; of colored minerals, 62; of permanganate of potash, 60.
Spectrum, comparison, 57; complementary, 56; continuous, 54; double, 57; incandescence, 56; line, 54; of malezeit sand, 62; single-banded, 61; solar, 54; two-banded, 61.
Spherical aberration, 4; distortion, 3.
Stage, mechanical, 28; micrometer, 37.
Stand of microscope, 1; for laboratory microscope, 28.
Standard distance (250 mm.) at which the virtual image is measured, 39.
Starch, examination of, 35.
Stokes and Angström's law of absorption spectra, 55.
Swaying of image, 21.
System, back, front, intermediate, of lenses, 3, 5.

T

Table of magnifications and valuation of ocular micrometer, 40; of tubelength and thickness of cover-glasses, 6. Of weights and measures (see inside of cover).
Terminology of objectives, 3.
Tester, cover-glass, 69.
Textile fibers, examination of, 35.
Thickness of cover-glass for non-adjustable objectives, table, 6.
Transmitted light, 16.
Transparent objects having curved outlines, relative position in microscopic preparations, 30.
Triplet, achromatic, 2.
Tripod, 2.
Tube-length, 5; for cover-glass adjustment, 24; importance of, 24; microscopical, 5; of various opticians, table, 6.
Turn-table, 71.

U—V—W

Unadjustable objectives, 5.
Unit of measure in micrometry, 41.
Valuation of ocular micrometer, 43.
Varying ocular micrometer valuation, 44.
Ward's eye-shade. 27.
Water immersion objective, 24.
Water for immersion objectives, 24; removal, 26.
Wave length, designation of, 58; scale of, 57.
Weights and measures, see inside of cover.
Wheat, examination of, 35.
Wollaston's camera lucida, 38, 48.
Working distance of simple microscope or objective, 15.

TABLE OF METRIC AND ENGLISH MEASURES.

The measures of length, volume and weight most frequently employed in microscopical and histological work are the following:

LENGTH.

1000 Microns (μ) = 1 Millimeter.
10 Millimeters (m.m.) = 1 Centimeter.
100 Centimeters (cm. or ctm.) = 1 Meter (unit of length).
1 μ = 0.000039 inch, $\frac{1}{25600}$th in. approximately.
1 cm. = 0.3937 in.
1 Meter = 39.3704 in.

VOLUME.

1000 Cubic centimeters (cc. or cctm.) or milliliters = 1 Liter (1000 grams of water) (unit of volume).
1 Fluid ounce (8 Fluidrachms) = 29.578 cc.

WEIGHT.

1000 Grams = 1 Kilogram (the weight of 1000 cc. or 1 liter of water).
1 Gram (unit of weight) = 15.432 Grains.
1 Kilogram = 2.204 Avoirdupois pounds.
1 Ounce Avoirdupois (437½ grains) = 28.349 Grams.
1 Ounce Troy or Apothecaries (480 grains) = 31.103 Grams.

TEMPERATURE.

To change Centigrade to Farenheit: (C. $\times \frac{9}{5}$) + 32 = F. For example, to find the equivalent of 10° Centigrade C. = 10° (10° $\times \frac{9}{5}$) + 32 = 50° F.
To change Farenheit to Centigrade: (F. − 32°) $\times \frac{5}{9}$ = C. For example, to reduce 50° Farenheit to Centigrade, F. = 50°, and (50° − 32°) $\times \frac{5}{9}$ = 10° C.; or − 40 Farenheit to Centigrade: F. = − 40° (− 40° − 32°) = − 72°, whence − 72° $\times \frac{5}{9}$ = − 40° C.

For the price of Microscopes and Microscopical supplies, the student is advised to obtain a catalogue of one or more of the Opticians named in the table of Tube-length and thickness of Cover-glass, (page 6), or of the following dealers in Microscopes and Microscopical supplies:

Eimer & Amend, 205–211 Third Avenue, New York.
The McIntosh Battery and Optical Co., 141–143 Wabash Avenue, Chicago, Ill.
James W. Queen & Co., 924 Chestnut St., Philadelphia, Pa.
Spencer & Smith, 250 Allen St., Buffalo, N. Y., (in place of H. R. Spencer & Co., Geneva).
W. H. Walmsley, 1022 Walnut St., Philadelphia, Pa.
Williams, Brown & Earl, 10th and Chestnut Sts., Philadelphia, Pa.
G. S. Woolman, 116 Fulton St., New York.

www.ingramcontent.com/pod-product-compliance
Lightning Source LLC
Chambersburg PA
CBHW031818230426
43669CB00009B/1181